T0201385

Single Channel Phase-Aware Signal Processing
in Speech Communication

Single Channel Phase-Aware Signal Processing in Speech Communication: Theory and Practice

Pejman Mowlaee

Josef Kulmer

Johannes Stahl

Florian Mayer

Graz University of Technology, Austria

Library of Congress Cataloging-in-Publication Data

Names: Mowlaee, Pejman, 1983- author. | Kulmer, Josef, author. | Stahl,
    Johannes, 1989- author. | Mayer, Florian, 1986- author.
Title: Single channel phase-aware signal processing in speech communication :
    theory and practice / [compiled and written by] Pejman Mowlaee, Josef
    Kulmer, Johannes Stahl, Florian Mayer.
Description: Chichester, UK ; Hoboken, NJ : John Wiley & Sons, Inc., 2016. |
    Includes bibliographical references and index.
Identifiers: LCCN 2016024931 (print) | LCCN 2016033469 (ebook) | ISBN
    9781119238812 (cloth) | ISBN 9781119238829 (pdf) | ISBN 9781119238836
    (epub)
Subjects: LCSH: Speech processing systems. | Signal processing. | Oral
    communication. | Phase modulation.
Classification: LCC TK7882.S65 S575 2016 (print) | LCC TK7882.S65 (ebook) |
    DDC 006.4/54—dc23
LC record available at https://lccn.loc.gov/2016024931

A catalogue record for this book is available from the British Library.

Cover Image: Gettyimages/lestyan4

Set in 10/12pt, WarnockPro by SPi Global, Chennai, India.
Printed and bound in Malaysia by Vivar Printing Sdn Bhd
10 9 8 7 6 5 4 3 2 1

Contents

6   Phase-Aware Speech Quality Estimation   *179*
        Pejman Mowlaee

7   Conclusion and Future Outlook   *210*
        Pejman Mowlaee

About the Authors

Dr Pejman Mowlaee (main author) Graz University of Technology, Graz, Austria

Pejman Mowlaee was born in Anzali, Iran. He received his BSc and MSc degrees in telecommunication engineering in Iran in 2005 and 2007. He received his PhD degree at Aalborg University, Denmark in 2010. From January 2011 to September 2012 he was a Marie Curie post-doctoral fellow for digital signal processing in audiology at Ruhr University Bochum, Germany. He is currently an assistant professor at the Speech Communication and Signal Processing (SPSC) Laboratory, Graz University of Technology, Austria.

Dr. Mowlaee has received several awards: young researcher's award for MSc study in 2005 and 2006, best MSc thesis award. His PhD work was supported by the Marie Curie EST-SIGNAL Fellowship during 2009–2010. He is a senior member of IEEE. He was an organizer of a special session and a tutorial session in 2014 and 2015. He was the editor for a special issue of the Elsevier journal *Speech Communication*, and is a project leader for the Austrian Science Fund.

Dipl. Ing. Josef Kulmer (co-author) Graz University of Technology, Graz, Austria

Josef Kulmer was born in Birkfeld, Austria, in 1985. He received the MSc degree from Graz University of Technology, Austria, in 2014. In 2014 he joined the Signal Processing and Speech Communication Laboratory at Graz University of Technology, where he is currently pursuing his PhD thesis in the field of signal processing.

Dipl. Ing. Johannes Stahl (co-author) Graz University of Technology, Graz, Austria

Johannes Stahl was born in Graz, Austria, in 1989. In 2009, he started studying electrical engineering and audio engineering at Graz University of Technology. In 2015, he received his Dipl.-Ing. (MSc) degree with distinction. In 2015 he joined the Signal Processing and Speech Communication Laboratory at Graz University of Technology, where he is currently pursuing his PhD thesis in the field of speech processing.

Florian Mayer (co-author) Graz University of Technology, Graz, Austria

Florian Mayer was born in Dobl, Austria, in 1986. In 2006, he started studying electrical engineering and audio engineering at Graz University of Technology, and received his Dipl.-Ing. (MSc) in 2015.

Preface

Purpose and scope

Speech communication technology has been intensively studied for more than a century since the invention of the telephone in 1876. Today's main target applications are acoustic human–machine communication, digital telephony, and digital hearing aids. Some detailed applications for speech communication, to name a few, are artificial bandwidth extension, speech enhancement, source separation, echo cancellation, speech synthesis, speaker recognition, automatic speech recognition, and speech coding. The signal processing methods used in the aforementioned applications are mostly focused on the short-time Fourier transform. While the Fourier transform spectrum contains both amplitude and phase parts, the phase spectrum has often been neglected or counted as unimportant. Since the spectral phase is typically wrapped due to its periodic nature, the main difficulty in phase processing is associated with extracting a continuous phase representation. In addition, compared to the spectral amplitude, it is a sophisticated task to model the spectral phase across frames.

This book is, in part, an outgrowth of five years of research conducted by the first author, which started with the publication of the first paper on "Phase Estimation for Signal Reconstruction in Single-Channel Source Separation" back in 2012. It is also a product of the research actively conducted in this area by all the authors at the *PhaseLab* research group. The fact that there is no text book on phase-aware signal processing for speech communication made it paramount to explain its fundamental principles. The need for such a book was even more pronounced as a follow-up to the success of a series of events organized/co-organized by myself, amongst them: a special session on "Phase Importance in Speech Processing Applications" at the International Conference on Spoken Language Processing (INTERSPEECH) 2014, a tutorial session on "Phase Estimation from Theory to Practice" at the International Conference on Spoken Language Processing (INTERSPEECH) 2015, and an editorial for a special issue on "phase-aware signal processing in speech communication" in *Speech Communication* (Elsevier, 2016), all receiving considerable attention from researchers from diverse speech processing fields. The intention of this book is to unify the recent individual advances made by researchers toward incorporating phase-aware signal processing methods into speech communication applications.

This book develops the tools and methodologies necessary to deal with phase-based signal processing and its application, in particular in single-channel speech processing. It is intended to provide its readers with solid fundamental tools and a detailed overview of the controversial insights regarding the importance and unimportance of phase in speech communication. Phase wrapping, exposed as the main difficulty for analyzing the spectral phase will be presented in detail, with solutions provided. Several useful representations derived from the phase spectrum will be presented. An in-depth analysis for the estimation of a signals' phase observed in noise together with an overview of existing methods will be given. The positive impact of phase-aware processing is demonstrated for three selected applications: speech enhancement, source separation, and speech quality estimation. Through several proof-of-concept examples and computer simulations, we demonstrate the importance and potential of phase processing in each application. Our hope is to provide a sufficient basis for researchers aiming at starting their research projects in different applications in speech communication with a special focus on phase processing.

Book outline

The book is divided into two parts and consists of seven chapters and an appendix. Part I (Chapters 1–3) gives an introduction to phase-based signal processing, providing the fundamentals and key concepts. Chapters 1–3 introduce an overview of the history of phase processing and reveal the phase importance/unimportance arguments (Chapter 1), the required definitions and tools for phase-based signal processing, such as phase unwrapping and abundant representations for spectral phase to make the phase spectrum more accessible (Chapter 2), and finally phase estimation fundamentals, limits potential, and its application to speech signals will be presented (Chapter 3).

Part II (Chapters 4–7) deals with three applications to demonstrate the benefit of phase processing in single-channel speech enhancement (Chapter 4), single-channel source separation (Chapter 5), and speech quality estimation (Chapter 6). Chapter 7 concludes the book and provides several future prospects to pursue. The appendix is dedicated to the implementations in MATLAB® collected as the *PhaseLab* toolbox in order to describe most of the implementations that reproduce the experiments included in the book.

Intended audience

The book is mainly targeted at researchers and graduate students with some background in signal processing theory and applications focused on speech signal processing. Although it is not primarily intended as a text book, the chapters may be used as supplementary material for a special-topics course at second-year graduate level. As an academic instrument, the book could be used to strengthen the understanding of the often mystical field of phase-aware signal processing and provides several interesting applications where phase knowledge is successfully incorporated. To get the maximal benefit from this book, the reader is expected to have a fundamental knowledge of digital signal processing, signals and systems, and statistical signal processing. For the sake of completeness, a summary of phase-based signal processing is provided in Chapter 2.

The book contains a detailed overview of phase processing and a collection of phase estimation methods. We hope that these provide a set of useful tools that will help new researchers entering the field of phase-aware signal processing and inspire them to solve problems related to phase processing. As the theory and practice are linked in speech communication applications, the book is supplemented by various examples and contains a number of MATLAB® experiments. The reader will find the MATLAB® implementations for the simulations presented in the book with some audio samples online at the following website:

https://www.spsc.tugraz.at/PhaseLab

These implementations are provided in a toolbox called *PhaseLab* which is explained in the appendix. The authors believe that each chapter of the book itself serves as a valuable resource and reference for researchers and students. The topics covered within the seven chapters cross-link with each other and contribute to the progress of the field of phase-aware signal processing for speech communication.

Acknowledgments

The intense collaboration in the year of working on this book project together with the three contributors, Josef Kulmer, Johannes Stahl, and Florian Mayer, was a unique experience and I would like to express my deepest gratitude for all their individual efforts. Apart from the very careful and insightful proofreads, their endless helpful discussions in improving the contents of the chapters and in our regular meetings led to a successful outcome that was only possible within such a great team. In particular, I would like to thank Johannes Stahl and Josef Kulmer for their full contribution in preparing Chapters 3 and 4. I would like to thank Florian Mayer for his valuable contribution in Chapter 5 and his endless efforts in preparing all the figures in the book.

Last, but not least, a number of people contributed in various ways and I would like to thank them: Prof. Gernot Kubin, Prof. Rainer Martin, Prof. Peter Vary, Prof. Bastian Kleijn, Prof. Tim Fingscheidt, and Dr. Christiane Antweiler for their enlightening discussions, for providing several helpful hints, and for sharing their experience with the first author. I would like to thank Dr. Thomas Drugman, Dr. Gilles Degottex, and Dr. Rahim Saeidi for their support regarding the experiments in Chapter 2. Special thanks go to Andreas Gaich for his support in preparing the results in Chapter 6. I am also thankful to several of my former Masters students who graduated at *PhaseLab* at TU Graz, Carlos Chacón, Anna Maly, and Mario Watanabe, for their valuable insights and outstanding support. I am grateful to Nasrin Ordoubazari, Fereydoun, Kamran, Solmaz, Hana, and Fatemeh Mowlaee, and the Almirdamad family who provided support and encouragement during this book project.

I would also like to thank the editorial team at John Wiley & Sons for their friendly assistance. Finally, I acknowledge the financial support from the Austrian Science Fund (FWF) project number P28070-N33.

Graz, Austria   *P. Mowlaee*
April 4, 2016

List of Symbols

$\lvert \cdot \rvert$	absolute value
$\angle$	angle
α	clean speech phase spectrum
α_0	tuning parameter for modified smoothed group delay
α_μ	mean value of the von Mises distribution
$\tilde{\alpha}$	perturbed clean speech phase
A	clean speech amplitude spectrum
A_h	amplitude of harmonic h
A_0	scale factor in the z-transform $X(z)$
$\hat{A}$	clean speech amplitude spectrum estimate
a_k	coefficients in the numerator polynomial of $X(z)$
arg$[\cdot]$	continuous phase function
ARG$[\cdot]$	principal value of phase
b_k	coefficients in the denumerator polynomial of $X(z)$
$\mathbf{B}_q$	basis matrix for the qth source in NMF
β_{DD}	smoothing parameter for decision-directed *a priori* SNR estimation
β_{SD}	smoothing parameter for the uncertainty in unvoiced speech
β	compression parameter of the parametric speech spectrum estimators
CG	coherent gain of a window function
$c(\cdot)$	compression function
$\Delta_B \phi$	baseband phase difference (BPD)
$\Delta\omega_{\mathrm{MW}}$	$-3\,\mathrm{dB}$ bandwidth of the window mainlobe
d_a	distance metric used in geometry-based phase estimator
d_{GDD}	GDD-based distance metric used in geometry-based phase estimator
$D_{-v}(\cdot)$	parabolic cylinder function
d	additive noise signal in time domain
d_w	additive noise along time with applied window function
Div	divergence measure
$D(k,l)$	DFT coefficient for noise
$D(e^{j\omega})$	DTFT of additive noise
D_w	DTFT of windowed noise frame
D_m	distance measure as squared error between two spectra
D_{MA}	mask approximation objective measure
D_{SA}	signal approximation objective measure

Δ_I	change in inconsistency
$\Delta\tau_x$	group delay deviation
$\Delta\phi$	phase deviation between the observation and the noisy signal
$\bar{d}_{\text{MCE}}$	cyclic mean phase error
$e^{(i)}$	remixing error in MISI for the ith iteration
$\mathbb{E}(\cdot)$	expected value operator
$\mathbb{E}(\cdot|\cdot)$	conditional expected value operator
ϵ	relative change of inconsistency
f_s	sampling frequency in Hz
f_0	fundamental frequency in Hz
$f_{0,q}$	fundamental frequency of qth source in mixture
ϕ_{dev}	phase deviation
ϕ	instantaneous phase from STFT
ϕ_{RPS}	relative phase shift
$_1F_1(\cdot,\cdot;\cdot)$	confluent hypergeometric function
G	gain function of a speech spectrum estimation scheme
$\mathcal{G}(\cdot)$	STFT(iSTFT($\cdot$))
γ	tuning parameter for modified smoothed group delay
γ_c	key adjustment parameter in CWF
γ_k	magnitude-squared coherence (MSC)
$\Gamma(\cdot)$	Gamma function
G_{PSF}	phase-sensitive filter
G_{CM}	complex mask filter
G_{CRM}	complex ratio mask filter
h	harmonic index
$\bar{h}$	desired harmonic
H	number of harmonics
H_0	hypothesis of no harmonic structure in the phase
H_1	hypothesis of harmonic structure in the phase
i	iteration index
I	maximum number of iterations
$I_v(\cdot)$	modified Bessel function of the first kind and order v
$\mathcal{I}(\cdot)$	inconsistency operator
ω_{IF}	discretized IF
Ω_q	confidence domain for the qth source in PPR approach
IBM	ideal binary mask
IRM	ideal ratio mask
IFD	instantaneous frequency deviation
j	imaginary unit
k	frequency index
κ	von Mises distribution concentration parameter
l	frame index
$l(\omega_k)$	integer-valued function used in time series phase unwrapping
L	number of frames
LC	local criterion used in IBM
Λ	phase spectrum compensation function
m	number of periods per window length

m_k	integer value as phase wrapping number
M	number of atoms used in NMF
N_{IUC}	number of zeros inside of the unit circle
N_{IOC}	number of zeros outside of the unit circle
μ	shape parameter of the parametric speech amplitude distribution
μ_h	circular mean parameter for the hth harmonic
μ_c	circular mean parameter of the von Mises distribution
μ_q	mean of the Gaussian distribution fitted to the qth source fundamental frequency
σ_q	standard deviation of Gaussian distribution fitted to the qth source fundamental frequency
n	sample index $n \in [0, N-1]$
n_0	instantaneous attack time
N_w	length of a window function
N	length of a frame
N_{DFT}	number of DFT points
$NMSE(\cdot)$	normalized mean square error
ω	normalized angular frequency
ω_0	fundamental radian frequency
ω_{IF}	instantaneous frequency (IF)
ω_h^k	closest sinusoid to bin k in STFTPI
o	tuning factor to scale mask in IRM
PC	phase change in Nashi's phase unwrapping method
PHINC	phase increment in Nashi's phase unwrapping method
P_{H_v}	voicing probability
ϕ_{lin}	linear phase along time
ϕ'	frequency derivative of phase
ϕ_h	phase value of harmonic h
$\hat{\phi}_h$	estimated phase value of harmonic h
PD	phase distortion
$p(\cdot)$	probability density function
ϕ_W	phase spectrum of the analysis window
q	source index in a mixture
Q	number of audio sources in a mixture
r_s	radial step size
ρ	Pearson's correlation coefficient
ρ_0	constant threshold used in ISSIR
r	phase randomization index
R	absolute value of noisy speech signal STFT
RPS	relative phase shift
$\mathcal{R}$	set of frames for von Mises parameter estimation
S	frame shift, hop size in samples
σ_x^2	speech variance
SI	speech intelligibility
SSR	signal-to-signal ratio (SSR)
SNR_a	SNR amplitude
SNR_p	SNR phase

$\mathrm{SNR}_{\mathrm{local}}$	local SNR
σ	normalized root-mean-square error
σ_{c}	circular variance
σ_d	noise variance
ψ	instantaneous harmonic phase
ψ_γ	objective function used in CWF
Ψ	unwrapped harmonic phase
$\hat{\psi}$	time–frequency smoothed harmonic phase
Ψ_q	activity domain used in ISSIR approach
t	time index
T	sampling period
τ_k	Kendall's tau
τ	group delay
τ_{m}	modified smoothed group delay function
τ_{p}	fixed threshold used in PPR approach
τ_{s}	smoothed group delay function
Φ	three-dimensional matrix for phase
UnRMSE	Unwrapped root mean square error
UnHPSNR	Unwrapped harmonic phase SNR
u	unvoiced speech signal components
U	unvoiced speech signal spectrum
Υ	anti-symmetry function used in phase spectrum compensation
ε_k	prediction error in adaptive numerical integration
V	vocal tract spectrum
$\mathbf{V}_q$	activation matrix for the qth source in NMF
$\angle V$	phase spectrum of the vocal tract (minimum phase)
$\mathcal{VM}(\mu,\kappa)$	von Mises distribution
ϑ	noisy speech phase spectrum
w	window function along time
W	frequency response for the window $w(n)$
W_k	band importance function for the kth frequency band
x	clean speech signal in time domain
x_{D}	deterministic speech component in time domain
$X_{\mathrm{D,W}}$	windowed deterministic speech spectrum
x_{SD}	stochastic–deterministic (SD) speech signal in time domain
x_0	zero-phase signal
x_l	signal frame
X	DFT of a clean speech signal
X_q	qth source DFT spectrum in a mixture
x_w	sequence along time with applied window function
X_w	DTFT of a windowed speech frame
X_{B}	baseband representation
X_{PS}	product spectrum
$X_{\Re}$	real part of the clean speech spectrum
$X_{\Im}$	imaginary part of the clean speech spectrum
ξ	*a priori* SNR
y	noisy speech in time domain

Y	noisy speech spectrum
$\tilde{Y}$	*a posteriori* mean for the stochastic–deterministic approach
Y_w	signal's modified STFT
y_w	modified signal
ζ	*a posteriori* SNR
z_m	mth zero in the z-plane

Part I

History, Theory and Concepts

1

Introduction: Phase Processing, History

Pejman Mowlaee

Graz University of Technology, Graz, Austria

1.1   Chapter Organization

This chapter provides the historical background on phase-aware signal processing. We will review the controversial viewpoints on this topic so that the chapter in particular addresses two fundamental questions:

- Is the spectral phase important?
- To what extent does the phase spectrum affect human auditory perception?

To answer the first question, the chapter covers the up-to-date literature on the significance of phase information in signal processing in general and speech or audio signal processing in particular. We provide examples of phase importance in diverse applications in speech communication. The wide diversity in the range of applications highlights the significance of phase information and the momentum developed in recent years to incorporate phase information in speech signal processing. To answer the second question, we will present several key experiments made by researchers in the literature, in order to examine the importance of the phase spectrum in signal processing. Throughout these experiments, we will examine the key statements made by the researchers in favor of or against phase importance. Finally, the structure of the book with regard to its chapters will be explained.

1.2   Conventional Speech Communication

Speech is the most common method of communication between humans. Technology moves toward incorporating more listening devices in assisted living, using digital signal processing solutions. These innovations show increasingly accurate and more and more robust performance, in particular in adverse noisy conditions. The latest advances in technology have brought new possibilities for voice-automated applications where acoustic human–machine communication is involved in the form of different speech

Single Channel Phase-Aware Signal Processing in Speech Communication: Theory and Practice, First Edition.
Pejman Mowlaee, Josef Kulmer, Johannes Stahl, and Florian Mayer.

Figure 1.1 Speech communication devices used in everyday life scenarios are expected to function robustly in adverse noisy conditions.

communication devices including digital telephony, digital hearing aids, and cochlear implants. The end user expects all these devices and applications to function robustly in adverse noise scenarios, such as driving in a car, inside a restaurant, in a factory, or other everyday-life situations (see Figure 1.1). These applications are required to perform robustly in order to maintain a certain quality of service, to guarantee a reliable speech communication experience. Digital processing of speech signals consists of several disciplines, including linguistics, psychoacoustics, physiology, and phonetics. Therefore, the design of a speech processing algorithm is a multi-disciplinary task which requires multiple criteria to be met.[1]

The desired clean speech signal is rarely accessible and is often observed only as a corrupted noisy version. There might also be some distortion due to failures in the communication channel introduced as acoustic echoes or room reverberation. Figure 1.2 shows an end-to-end speech communication consisting of the different blocks required to mitigate the detrimental impacts causing impairment to the desired speech signal. Some conventional blocks are de-reverberation, noise reduction including single/multi-channel signal enhancement/separation, artificial bandwidth extension, speech coding/decoding, near-end listening enhancement, and acoustic

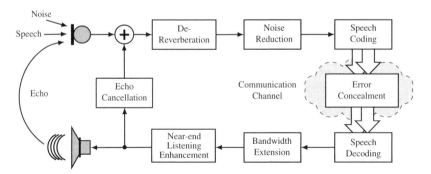

Figure 1.2 Block diagram for speech communication from transmitter (microphone) to receiver end (loudspeaker) composed of a chain of blocks: beamforming, echo cancellation, de-reverberation, noise reduction, speech coding, channel coding, speech decoding, artificial bandwidth extension, near-end listening enhancement.

---

1 See Chapter 6 for speech quality estimation used for performance evaluation purposes.

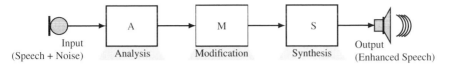

Figure 1.3 Block diagram of the processing chain in speech communication applications: analysis–modification–synthesis.

echo cancellation. Depending on the target application, several other blocks might be considered, including speech synthesis, speaker verification, or automatic speech recognition (ASR), where the aim is the classification of the speech or the speaker, e.g., for forensics or security purposes.[2]

Independent of which speech application is of interest, the underlying signal processing technique falls into the unified framework of an *analysis–modification–synthesis* (AMS) chain, as shown in Figure 1.3. The *short-time Fourier transform* (STFT) is frequently used to analyze and process the speech signal framewise. In speech signal processing, a frame length between 20 to 40 milliseconds is quite common, together with a *Hamming* window (Picone 1993; Huang *et al.* 2001). The modification stage involves modifying the spectral amplitude, spectral phase, or both. The conventional methods attempt to modify the spectral amplitude only and a large variety of literature has been devoted to deriving improved filters to modify the spectral amplitude of speech signals. For the synthesis, the inverse operator to the analysis is applied, which is typically the *inverse short-time Fourier transform* (iSTFT).

In contrast to the extensive literature for spectral amplitude modification in digital speech transmission (see, e.g., Vary and Martin 2006), fewer works have been dedicated to phase processing or incorporating phase information in speech processing applications. The main reason is that the Fourier phase spectrum contains no useful structure due to its cyclic wrapping.[3] This results in difficulties in processing the spectral phase or how it is interpreted, thus rendering the phase spectrum largely inaccessible.

It took researchers a century to revisit the importance of phase information in different speech processing applications. Only recently has research been conducted toward incorporating phase information at the modification stage in various speech processing applications. We discuss three such applications in the second part of the book, Chapters 4–6. In the following, we present a historical review to highlight the reasons why the spectral phase was assumed to be unimportant or important. We justify the key statements by various researchers throughout, explaining their experiments. The current chapter continues to address the following two questions:

- **Is the phase irrelevant/unimportant/important?** Why has phase been a controversial topic, believed by some researchers irrelevant or unimportant and by some others important and perceptually relevant?
- **If the phase is important, to what extent?** When is it perceptually relevant and when not?

---

2 The diversity of such speech applications and the recent advances towards incorporating phase-aware signal processing solutions are reviewed in Section 1.4; two selected exemplary applications are presented in the second part of the book in Chapters 4–6.

3 See Chapter 2 for more details on phase wrapping and several available solutions.

To answer the first question, in Section 1.3 we present a detailed historical review of the studies dedicated to the importance of the phase spectrum in speech signal processing. In Section 1.4 we provide examples of the phase information that has been incorporated in different speech communication applications. To answer the second question, in Section 1.6 we demonstrate several experiments to investigate the importance of phase in human perception.

1.3 Historical Overview of the Importance or Unimportance of Phase

The processing and treatment of the phase spectrum in speech signal processing applications has largely been neglected in the past. The first insights with regard to phase perception date back to 1843 when Georg Simon Ohm (Ohm 1843) and Hermann von Helmholtz (Helmholtz 1912) concluded from their experiments that human ears are *insensitive to phase*. Helmholtz studied the concept of the Fourier series analysis of periodic signals. He further visualized the cochlea in the human ear as a spectral analyzer. Helmholtz claimed that the magnitude of the frequency components are the sole factors in the perception of musical tones. He suggested that humans perceive musical sounds in the form of harmonic tones, concluding with the following statement: "the perceived quality of sounds does not depend on the phases but on the existence of the underlying particular frequencies in the signal."

Seebeck used sirens to produce periodic stimuli with several pulses irregularly spaced within one period (Turner 1977). He identified that perception of a tone is not solely dependent on the relative strength of the underlying harmonics in the signal, which contradicted the earlier observations made by Helmholtz. He found that the rate of the impulses involved in the sound production contributes to the ultimate perceptual quality of sound. This important observation was to highlight the importance of pitch information in a signal, and led to the *Ohm−Seebeck dispute* (Turner 1977).

Weiss *et al.* (1974) demonstrated that rapid fluctuation in the phases of the underlying sinusoids in speech leads to significant degradation in the speech quality. They further developed a process to improve the *signal-to-noise ratio* (SNR) of speech corrupted with some wideband noise. In 1975, Schroeder presented a thorough review of hearing models and studied the importance of the phase spectrum in human perception (Schroeder 1975). In particular, he contributed to the effects of monaural phase by addressing the fundamental question of to what extent the human auditory system decodes the phase spectrum of a signal. He described the perceptual effect of changing phases in complex harmonic signals composed of up to 31 components. Schroeder, in his work entitled "New results concerning monaural phase sensitivity" (Schroeder 1959), demonstrated a phenomenon called Schroeder phase in which by just modifying the individual phase components of two signals of identical envelopes, it is possible to produce strong varying pitch perception, e.g., when playing melodies.

Later, several simple demonstrations were made to oppose the earlier belief that the *human ear is phase-deaf*. This series of studies included Bilsen (1973) and Carlson *et al.* (1979). In contrast to the earlier conclusions made by Helmholtz, they demonstrated perceptible changes in the clarity of pitch (Bilsen 1973) or timbre

(Plomp and Steeneken 1969) due to phase changes in the signal. Later, Schroeder and Strube (1986) demonstrated the dependence of the short-time spectral amplitude on phase changes. They justified this hypothesis by conducting experiments showing the possibility of constructing intelligible voiced speech signals that have a flat spectral amplitude. As another example, Ni and Huo (2007) presented a statistical interpretation of the phase information in signal and image reconstruction. Throughout their experiments, they demonstrated that important features of a signal are preserved in its phase spectrum. All these references and the observations therein commonly show that the simple models of hearing proposed earlier by Ohm and endorsed by Helmholtz must fail for both musical tones and vowels. However, these studies do not provide details on how and in which range the human ear is sensitive to phase distortion.

Lipshitz *et al.* (1982) studied the audibility of phase distortion in audio systems. They concluded that even quite small to mid-range non-linear phase distortions could become audible, an effect better perceived through headphones rather than loudspeakers. Throughout listening tests they showed that these mid-range phase distortions are audible in the case of combined sinusoids, which is common in acoustic signals. Patterson (1987) conducted experiments to verify the capability of the human auditory system to discriminate the changes in the phase spectra of wideband periodic signals. A series of phase changes was applied to simulate envelope modifications of waves at the output of auditory filters. The amount of these local phase changes was shown to be strongly related to the spectral location, the repetition range, and the intensity of the underlying signal. To explain this observation, a highly simplified model of the cochlea was suggested where each filter was modeled as some units recording the time occurrence of large peaks in the filter output. Through the experiments, Patterson concluded that the differences in the phase spectra become imperceptible only when the fundamental frequency exceeds 800 Hz.

In another study (Patterson 1994), Patterson proved that phase information is important for auditory perception and proposed a model that simulated the auditory perception property, whereby phase perception relies on phase-locking in the firing of the neurons sitting in the cochlea. In the case of a single tone, the neuronal firing rate of the hair cells represents one pulse per period of the underlying sinusoid corresponding to its phase. As a result, Patterson reported that depending on the synchrony of neural firing between different frequencies, different sounds could be perceived. Patterson (1987) also reported that phase perception is limited to low pitch sounds as the phase-locking effect is lost for high frequencies. In particular, Patterson suggested that humans are *phase-deaf* for sounds with repetition rates higher than 400 Hz while sensitive for rates below 200 Hz.

Liu *et al.* (1997) studied the significance of phase in the perception of intervocalic stop sounds. In his study he conducted listening tests with magnitude-only and phase-only reconstructions. It was observed that the phase spectrum takes precedence over the magnitude spectrum when long analysis windows are chosen. Laitinen *et al.* (2013) studied the sensitivity of human hearing to changes in the phase spectrum for a synthetic harmonic complex signal. Their test confirmed that humans are most sensitive to harmonic complex signals where the phases of the underlying harmonics are related in a certain way. Examples of such signals are trumpet sounds and speech produced by a physiological sound source. They further presented an auditory model to describe the effects caused by modifying the phase spectrum (Laitinen *et al.* 2013). Based on this auditory model they further developed a measure to detect whether humans perceive a certain change in the phase spectrum.

In many classical signal processing applications, we are given a signal spectrum that is incomplete in the sense that the phase and magnitude spectra are not necessarily both accurately observed. Therefore, it is desired to reproduce the original signal from such an incomplete/distorted spectrum. Reproduction from such a deficient spectrum is termed *retrieval of the remaining information*. In the 1980s, several researchers investigated how to estimate a time-domain signal only from its STFT magnitude information. The problem arises in many applications, including time-scale modification, speech decoding, speech synthesis, speech enhancement, and source separation, where no access to the original phase spectrum is possible, or the phase of the received signal is distorted or noise-corrupted.

Oppenheim *et al.* (1979) explored the possibility of reconstructing a digital signal from its partial Fourier information. In particular, they considered two extreme scenarios: magnitude-only or phase-only signal reconstruction. Oppenheim *et al.* clarified the importance of phase information in the sense that under certain circumstances, e.g., when the signal is of finite length, then the phase-only information suffices to restore the signal up to a scaling factor (Oppenheim and Lim 1981). They carried out experiments with long-term phase information and demonstrated that such a representation contains important information about the underlying speech signal. They further exemplified their proposed algorithms for image and audio signals. The importance of phase-only signal reconstruction was demonstrated on human perception in terms of the quality of the reconstructed signals. They showed that phase contains a great deal of information about a signal and exemplified this fact through examples in image and speech signals where phase-only reconstruction could retain the intelligibility (for speech) or visualization (for images) of the original signal, respectively. Finally, they proposed algorithms to reconstruct the signal from phase information (Oppenheim *et al.* 1983). They further reported results on the exact reconstruction of a signal from its signed-magnitude-only or phase-only spectrum. Finally, Quatieri (1979) presented iterative algorithms for the retrieval of phase information from magnitude-only observation of a signal. He further developed algorithms to extract the magnitude from the phase spectrum under the causality and the finite duration constraint, imposed on the underlying signal of interest. A detailed description of these methods will be presented in Chapter 3.

As other examples, techniques for signal reconstruction using phase-only or magnitude-only STFT spectra were reported in Hayes *et al.* (1980) and Quatieri and Oppenheim (1981). The same authors further studied the conditions under which a signal can be reconstructed from one-bit-phase-only information and successfully applied them for image reconstruction (Hayes *et al.* 1980). In 1981, Quatieri *et al.* proposed an iterative minimum phase signal reconstruction algorithm from phase or magnitude (Quatieri and Oppenheim 1981). Several studies were later dedicated to estimating a time-domain signal from the magnitude information of the short-time Fourier transform, as well as on how to use the phase-only information to achieve a unique reconstruction of a time-domain signal (Hayes *et al.* 1980; Quatieri and Oppenheim 1981; Griffin and Lim 1984). Espy and Lim (1982) experimentally studied the effects of additive noise in the given phase on signal reconstruction from the

Fourier transform phase. The accuracy of the reconstructed signal was found to be quite sensitive to the accuracy of the spectral phase information.

As a highly important contribution to the field, Griffin and Lim (1984) formulated the problem of iterative signal reconstruction from the STFT magnitude in an optimal *minimum mean square error* (MMSE) framework. They proposed an iterative signal reconstruction algorithm that will be discussed in Chapter 3, together with some extensions used to estimate the phase information given a modified-magnitude STFT observation. Guozhen and Morris (1989) examined the role of phase in digital signal spectra and proposed an iterative technique to reconstruct images from their phase information. These works disproved the earlier observations of Helmholtz and Ohm, showing the potential importance of spectral phase information signal processing in general and in audio/speech processing in particular.

Paliwal *et al.* addressed the importance of the phase spectrum in a series of publications. A general overview of phase importance in speech processing was provided in Alsteris and Paliwal (2007), and in more detail for certain speech applications, including human listening (Paliwal and Alsteris 2005, 2003), speech enhancement (Paliwal *et al.* 2011), automatic speech recognition (Alsteris and Paliwal 2004), speech intelligibility (Alsteris and Paliwal 2006), and speech analysis (Stark and Paliwal 2008, 2009). Alsteris and Paliwal (2007) presented an overview of STFT phase in speech processing applications and reviewed experimental results for short-time phase spectrum usage in speech processing. This work demonstrated the usefulness of the phase spectrum for automatic speech recognition, human listening, and speech intelligibility. It was observed that the spectral phase information helps in improving the accuracy of automatic speech recognizers (Alsteris and Paliwal 2007).

Given this historical overview on the unimportance/importance of phase, in the following section we will present a focused literature review on how phase has successfully been taken into account in different speech processing applications.

1.4 Importance of Phase in Speech Processing

We discuss the phase-aware signal processing carried out in the following speech communication applications:

- speech enhancement
- speech watermarking
- speech coding
- artificial bandwidth extension
- automatic speech recognition
- speaker recognition
- speech synthesis

1.4.1 Speech Enhancement

In the following, a literature review will be provided with regard to the importance of phase in speech enhancement. Details of phase-aware speech enhancement and source separation will be explained in Chapters 4 and 5, respectively.

1.4.1.1 Unimportance of Phase in Speech Enhancement

In the 1980s, two important experiments were conducted that highlighted the beliefs about the importance or unimportance of phase information in speech enhancement. The first experiment was reported by Wang and Lim in 1982 as "The unimportance of phase in speech enhancement" (Wang and Lim 1982). They performed listening experiments where the spectral amplitude of noise-corrupted speech at a given SNR was combined with the spectral phase information distorted with some controlled added noise. The participants in the listening experiments were asked to identify the equivalent SNR for the reference speech signal in which the perceived quality of the test stimuli and the reference signals were graded equally. They reported that "a more accurate phase estimate than the degraded phase will not significantly improve the equivalent SNR." The results achieved by Wang and Lim showed that when a less distorted phase is combined with the noisy spectral amplitude, an improvement in SNR of 1 dB or less is achievable. Therefore, they concluded with the statement that improving the accuracy of the spectral phase information is not critical for the speech enhancement application (Wang and Lim 1982). Details of Wang and Lim's experiment will be presented in Section 1.6.1.

1.4.1.2 Effects of Phase Modification in Speech Signals

In 1985 it was Vary who studied the relevance of the short-time spectral phase and the effect of phase modification on the perceived quality of the reconstructed speech signal (Vary 1985). He conducted a series of experiments within the framework shown in Figure 1.4. The phase modification experiment is visualized in Figure 1.6, with the details explained, in Section 1.6.2.

Through his experiments Vary observed that if the original clean phase is replaced by zero-phase, the synthesized speech is perceived to sound completely voiced and *monotonous* (perceived as constant pitch). On the other hand, if the clean phase is replaced with a randomly distributed phase between $-\pi$ and π, the reconstructed speech outcome sounds rough and completely unvoiced. Finally, when a controlled level of randomness is added to the clean phase, Vary showed that the roughness due to phase modification in speech enhancement is only perceptible for signal components with instantaneous SNRs below 6 dB. In 1985, Vary also reported that as long as the maximum noise amplitude is below a certain threshold, no perceptual distortion in phase is recognized. The limit of perception of phase deviation (defined as the difference between the clean and noisy spectral phase) was found to be between $\frac{\pi}{8}$ and $\frac{\pi}{4}$. For details of the experiment, see Section 1.6.2.

1.4.1.3 Phase Spectrum Compensation

Paliwal *et al.* (2011) studied the importance of phase information for speech enhancement and proposed the *phase spectrum compensation* (PSC) method using the conjugate symmetricity property of the *discrete Fourier transform* (DFT) of a real signal. The PSC

Figure 1.4 The experimental setup in Vary (1985) comprised three stages: a spectral analyzer, either a polyphase network (PPN) or fast Fourier transform (FFT), followed by an adaptive processor (amplitude/phase modification) and a spectral synthesizer.

method was originally proposed in Stark *et al.* (2008) and Wojcicki *et al.* (2008), and will be covered in Section 1.6.4, where its success in speech enhancement will be demonstrated.

1.4.1.4 Phase Importance for Improved Signal Reconstruction

Kazama in 2010 studied the importance of phase for speech intelligibility. He concluded that for an overlap of 50% the phase becomes important for long enough segments (longer than 128 ms) or short enough segments (shorter than 2 ms). This observation was in agreement with those reported earlier by Wang and Lim (Wang and Lim 1982). Paliwal *et al.* (2011) showed significant improved performance by the STFT phase spectrum in terms of perceived quality predicted by the ITU standard perceptual evaluation of speech quality (PESQ; Rix *et al.* 2001), up to 0.2 points in white noise and 0 dB signal-to-noise ratio. It is important to note that in Paliwal's experiments the overlap was 87.5%, in contrast to the experiments in Kazama *et al.* (2010) or Wang and Lim (1982), as Paliwal *et al.* zero-padded the speech segments before applying the Fourier analysis.

Gerkmann *et al.* presented an overview of phase-aware single-channel speech enhancement applications (Gerkmann *et al.* 2015). Mowlaee and Kulmer (2015a) presented an overview focused on phase estimation for single-channel speech enhancement. Throughout their review, they demonstrated the potential and limits of phase estimation methods evaluated with a comparative study. In Mowlaee and Kulmer (2015b), the same authors proposed SNR-based phase estimation. Mayer and Mowlaee presented an overview of phase estimation methods used for single-channel source separation (Mayer and Mowlaee 2015). The main contribution of Mowlaee and Kulmer (2015a), Mowlaee and Kulmer (2015b), and Mayer and Mowlaee (2015) was on phase estimation in noise and incorporating it for improved signal reconstruction. Maly and Mowlaee (2016) studied the importance of harmonic phase modification and reported that the linear phase combined with an unwrapped phase suffices for improved signal reconstruction in speech enhancement.

The positive impact of an enhanced phase for signal reconstruction will be discussed in depth in Chapter 3, while further details regarding phase-aware speech enhancement will be explained in Chapter 4.

1.4.2 Speech Watermarking

In watermarking, the goal is to find a robust way to hide data (the watermark) inside a host signal such that it is not revealed to unauthorized users. While the main focus of audio watermarking has been dedicated to the spectral amplitude, the phase spectrum of unvoiced speech has been considered to be a good candidate for watermark signals. The rationale behind this choice is the fact that for the unvoiced phase spectrum the human auditory system is not capable of distinguishing different realizations of the Gaussian excitation process as long as the temporal-spectral envelope characteristics of the speech are not changed. A data hiding capacity of up to 540 bits/s (Kubin *et al.* 1993; Hofbauer *et al.* 2009), robust to nonlinear phase and bandpass filtering attacks, was achieved.

Dong *et al.* (2004) proposed a data hiding algorithm relying on the phase modulation concept. The idea was to modify the relative phases of the underlying spectral components in a given audio signal. For the purpose of data hiding, they applied the quantization index modulation principle. Here, a variable set of phase quantization steps is used, yielding a data hiding capacity of 20 kbit/min. Liew and Armand (2007) proposed to embed watermark data in the phase of some randomly chosen frequencies for low amplitude spectral components. The method relies on the insensitivity of the human auditory system to absolute phase values. In order to improve the robustness of their method, they also proposed a variable framing strategy. As another example, Kuo *et al.* (2002) proposed a watermarking scheme relying on the perceptual insignificance of long-term multi-band phase modulation. The method was reported for a 20 bits/s data hiding capacity and robust to coding algorithms. Finally, Hernaez *et al.* (2014) proposed a watermarking method relying on harmonic phase encoding. The idea was to preserve the harmonic structure in phase captured by the *relative phase shift* (RPS;[4] Saratxaga *et al.* 2012), which has been demonstrated as a critical feature for high quality synthesized speech (de Leon *et al.* 2011; Sanchez *et al.* 2015). The method was reported for a data hiding capacity of one bit per frame, and even showed robustness to coding algorithms like MP3.

1.4.3 Speech Coding

In digital coding of speech signals, it is often recommended to assign more bits for spectral amplitude/envelope information, rather than for the phase spectrum. This is mainly due to the higher perceptual relevance of the spectral amplitude compared to phase information. However, the harmonic phase information has also been reported to affect the naturalness and perceived quality of coded speech, in particular at low to mid rates (McAulay and Quatieri 1986).

Kim (2000) studied the importance of phase quantization in terms of the perceptual artifacts introduced by the harmonic coder; he evaluated the coders in the sense of their perceptual importance regarding the harmonic phase. Kim reported that phase is perceptually relevant, in particular at frame boundaries of two adjacent frames as soon as the fundamental frequency changes. These findings by Kim confirmed the earlier observations by Pobloth and Kleijn (1999) and Skoglund *et al.* (1997) which demonstrated the importance of phase in perception. In Kim (2003), the *just-noticeable difference* (JND; Zwicker and Fastl 2007) was measured for the phase distortion of one harmonic versus different fundamental frequencies. A mathematical model of the JND was also proposed as a quantization method. Such a model made it possible to assign bits to the perceptually most important part of the phase spectrum. Eventually their approach led to a coded speech signal that was perceptually close to the uncoded one.

Honda (1989) proposed analysis-synthesis using a phase equalization concept. Honda's experiments showed that the synthesized speech signal using phase-equalized excitation suffers from "buzzyness" sound characteristics, hence showing the effectiveness of phase information for high quality coded speech. They proposed a new short-time phase representation relying on group delay for speech coding.[5] It was demonstrated that the new representation preserved the glottal shape of the speech signal's group delay, found to be important for auditory perception. Their proposed representation is useful for reducing the bit rate as well as speech morphing effects. They reported high quality speech coding at bit rates lower than 1.8 kb/s. It was

---

4 The relative phase shift (RPS) representation will be explained in detail in Chapter 2.
5 In Chapter 2, details of the group delay as a useful phase representation will be presented.

also demonstrated that simply using zero or minimum phase for speech synthesis significantly damages the perceptual quality.

In low-rate harmonic coders, the phase information is often not transmitted, but regenerated at the decoder for speech synthesis. This can be accomplished by phase estimation using the minimum-phase assumption (Brandstein *et al.* 1990) and predicting the harmonic phase based on the frequencies in the previous and current frames (Kleijn and Paliwal 1995), similar to the principle used in phase vocoders (Dolson 1986). In Pobloth and Kleijn (1999), it was shown that the human perception capacity towards phase is considerable, and therefore existing speech coders introduce some artifacts, in particular for low-pitched voice. In Pobloth and Kleijn (2003), the authors studied the relevance of a squared error measure defined between the original and phase-distorted signal. While for small values of phase distortion the squared error correlates well with the perceptual measure, for large values in the squared error it does not indicate any correlation.

Kim showed in his studies (Kim 2000, 2003) that the phase information below a certain frequency in a harmonic signal is not perceptually relevant, as they result in a lower JND compared to high frequencies. This certain frequency was termed the *critical phase frequency*. For example, for a fundamental frequency of $f_0 = 100$ Hz, a critical phase frequency index of seven was obtained through listening tests. Finally, in Yu and Chan (2004), it was shown that encoding the phase of harmonics above or equal to the critical phase frequency suffices to preserve the perceptual quality of coded speech.

In low bit-rate speech coding, the predictability of lowband harmonics across frames has been taken into account in order to minimize the number of bits assigned for harmonic phase representation. In particular, in *mixed excitation linear prediction* (MELP) and *code-excited linear prediction* (CELP) coders, the phase information is transmitted only at peaks, whereas at the decoder, the phase information is synthesized using interpolation techniques. For example, McAualay and Quatieri suggested cubic interpolation to provide a smooth spectral phase to synthesize speech signals (McAulay and Quatieri 1986). In order to maintain a transparent speech quality, they report that an appropriate phase for harmonics is required. In Agiomyrgiannakis and Stylianou (2009), phase dispersion was used to train codebooks for speech coding.

Finally, it is important to note that the correct alignment of sinusoidal trajectories regarding their phase coherence has been well studied in the literature of harmonic speech coding. Such a proper alignment is highly important for the sake of high quality synthesized speech. Following this point of view, several phase interpolation techniques can be found in the literature. McAulay and Quatieri (1986) proposed cubic interpolation of the phase values across frames, with the assumption of minimum phase for the phase contribution by the vocal tract filter. In Pobloth (2004), the importance of the dispersion phase was taken into account when reconstructing speech using a harmonic coder. Agiomyrgiannakis (2015) presented an overview on phase smoothing methods used for harmonic speech synthesis where he proposed phase-locked pitch-synchronous and quadratic phase smoothing ideas.

1.4.4 Artificial Bandwidth Extension

The impairment of speech during its transmission as shown in Figure 1.2 is not limited only to noise or echoes. In fact, the traditional telephone network used for digital speech transmission is a bottleneck, as it limits the effective bandwidth of the transmitted speech to 300–3400 Hz, termed narrowband sound quality. This delivers poor speech quality when compared to the more pleasant *wideband speech* experience defined with a frequency range of 50–7000 Hz. To circumvent the band-limited nature of telephony speech, artificial bandwidth extension techniques have been successfully used in speech codecs leading to several wideband codec standards (see Iser *et al.* 2008 for an overview). The information loss at low and high frequency bands results in losing certain speaker characteristics, which contributes to poor speaker identification, in particular muffled high frequencies of fricative sounds (e.g., /s/,/f/). The missing low frequencies lead to reduced loudness (also termed *thin voice*; Jax 2002), degrading the perceived quality. Using artificial bandwidth extension methods enables improved perceived quality and speech intelligibility.

The state-of-the-art principle used for bandwidth extension is to apply the well-known source-filter model (for more details see Chapter 2 in Vary and Martin 2006). Then the bandwidth extension is divided into two subtasks: extension of the narrowband envelope, and extension of the excitation signal. The source-filter model does not explicitly take into account any phase contribution in the excitation, but the minimum phase introduced by the spectral envelope. Only a few researchers took into account some phase modification to improve the quality of the estimated wideband speech. For example, to mitigate the undesired impact due to applying strict periodic excitation, Nilsson and Kleijn (2001) proposed the combination of spectrum folding with adding white noise. Later, Heese *et al.* (2010) proposed adding filtered white noise at missing high band frequencies, which showed improved naturalness. Geiser and Vary (2013) showed that adding a randomized phase at missing high frequency bands results in improved bandwidth extension performance. Pulakka *et al.* (2012) applied a linear phase model to reconstruct the phase for low bandwidth extension of narrowband speech. Peharz *et al.* (2014) used the Griffin and Lim iterative signal reconstruction algorithm to estimate the phase of the missing high frequencies.[6]

1.4.5 Speech Synthesis

Speech synthesis techniques mostly rely on concatenating phonemes (Beutnagel *et al.* 1999). The most important observation is that the linear phase part mismatches at the concatenation boundaries, resulting in undesired audible clicks in the synthesis outcome. To mitigate this problem, Stylianou (2001) defined a reference point called the *center of gravity*, approximated as the phase of the first harmonic.

Another phase-aware contribution for speech synthesis is the band aperiodicity (BAP) method (Zen *et al.* 2006). The idea is to circumvent the insufficiency of the linear phase plus minimum phase to achieve a high perceived synthesized quality. Noise is added in selected frequency bands in order to reduce the highly unnatural correlation resulting from employing only linear phase and minimum phase. Later, Maia *et al.* (2012) proposed complex cepstrum factorization in order to jointly model the amplitude–phase spectra of a speech signal. Further, Maia and Stylianou (2014) demonstrated that factorizing speech to minimum phase and all-pass filters results in the best synthesized speech quality. Agiomyrgiannakis (2015) proposed an amplitude modulation (AM)–frequency modulation (FM) sinusoidal model for parametric speech synthesis. The high quality

---

6 The Griffin and Lim iterative signal reconstruction algorithm will be presented in detail in Chapter 3.

synthesized speech results revealed the fact that a proper phase model is capable of boosting the quality of the synthetic speech in statistical parametric speech synthesis systems. Finally, Maia and Stynlianou (2016) proposed an iterative method to estimate the phase using the complex cepstrum representation of speech. The method overcame two major issues known as bottlenecks of the conventional complex cepstral analysis methods: (i) the need for an accurate *glottal closure instant* (GCI) estimate, and (ii) the need for phase unwrapping.[7] Their proposed method was reported to perform better than methods relying on GCI markings and phase unwrapping at a lower computational complexity.

1.4.6 Speech/Speaker Recognition

Phase was also reported to be important for human speech recognition. Aarabi *et al.* (2006) presented an overview on phase-based speech processing mainly focused on robust speech enhancement (Shi *et al.* 2007) and improved automatic speech recognition (Guangji *et al.* 2006). They reported improved digit recognition accuracy (where tests were conducted by several speakers) using the difference between the Fourier transform phase of two microphones. The central idea was motivated by the fact that the phase difference for a clean non-reverberant speech signal is equal after delay compensation. Based on the phase difference information, they derived a mask to be applied to the noisy signal for enhancement purposes. They demonstrated that the signal-to-noise ratio at each frame and frequency is a function of the phase error (Shi *et al.* 2007). This was incorporated to derive a perceptually motivated phase-error filter for speech enhancement (Aarabi and Shi 2004).

Most of the *automatic speech recognition* (ASR) or speaker recognition systems are built upon short-term spectro-temporal feature representations, typically calculated from the amplitude spectrum (Davis and Mermelstein 1980). However, new features derived from the spectral phase have become a popular approach and have been reported to be useful to improve the overall recognition accuracy. In Chapter 2, we will present the group delay and its modified extensions, for example the *modified group delay* (MGD) function (Hegde *et al.* 2004; Loweimi and Ahadi 2011), as some useful phase representations. McCowan *et al.* (2011) proposed using the delta-phase spectrum as a feature for speaker recognition. Kleinschmidt *et al.* (2011) proposed complex spectral subtraction for robust automatic speech recognition. Loweimi *et al.* (2015) proposed a source-filter decomposition based on phase-domain processing. In particular, they showed that the imperfect source-filter decomposition in log-magnitude spectra can be better performed by trend and fluctuation analysis of the minimum phase spectrum of speech. The usefulness of source-filter decomposition in the phase domain was demonstrated in ASR with absolute improvement in comparison to the conventional Mel-frequency cepstral coefficient (MFCC) features.

Processing the raw time signal instead of extracting features like MFCC has been presented as an alternative signal processing approach for ASR (Tüske *et al.* 2014; Golik *et al.* 2015). Golik *et al.* (2015) trained a *convolutional neural network* (CNN) with raw time signals to learn acoustic models, which is contrary to conventional feature extraction routines relying on filter banks and an energy feature extraction stage. The CNN is capable of learning the critical band energy filters distributed non-linearly, similar to

---

7 See Chapter 2 for a detailed review of phase unwrapping methods.

the Mel filterbank. Using raw time signals together with a CNN, a WER comparable to MFCC-based ASR using fully connected layers was reported. The position of the learned filters in time is distributed uniformly within the context (Tüske *et al.* 2014). This is in contrast to the MFCC-based features, where the position of the filters in time is bounded to the center of the stacked samples of the signal. As these time offsets or shifts are only present in the phase spectrum, the authors concluded that a CNN is capable of learning different filters at different segments of the signal. By providing raw time signals to the neural network, the acoustic models can learn non-stationarity patterns with a reduced loss of information due to the processing used in common feature extraction.

The consistency property of the speech spectral phase has been used for anti-spoofing for speaker recognition. This characteristic has been used to develop a counter measure for synthetic speech detection (de Leon *et al.* 2011; Wu *et al.* 2015). Given a phase-derived stream of features, it is possible to combine them directly with the amplitude features (Schlüter and Ney 2001) or at their score-levels (Murty and Yegnanarayana 2006; Nakagawa *et al.* 2006; Wang *et al.* 2010; Nakagawa *et al.* 2012). Finally, to unify the feature extraction of amplitude and phase, the Hilbert transform has been proposed for speaker recognition (Murty and Yegnanarayana 2006; Sadjadi and Hansen 2011) and automatic speech recognition (Loweimi *et al.* 2013). Xiao *et al.* (2015) proposed an automatic speaker verification spoofing system where seven features including two magnitude-based and five phase-based features were employed. As the phase spectrum does not contain stable patterns for spoofing speech detection due to its cyclic wrapping (see Chapter 2), Xiao *et al.* employed several phase-based features: group delay, modified group delay, instantaneous frequency derivative, baseband phase difference, and pitch synchronous phase features. Finally, Tian *et al.* (2015) studied the importance of phase-based features for the synthetic speech detection problem. They considered an instantaneous frequency derivative, the modified group delay spectrum, and the modified group delay spectrum cepstral coefficients. These phase-derived features were reported to be complementary to the conventional MFCC features.

1.5 Structure of the Book

The goal of this book is to inform the reader about the fundamental problems of phase-aware signal processing for speech applications. The book presents the theoretical aspects and the various advances made to incorporate the phase spectrum in different speech communication examples. The book consists of seven chapters and is divided into two parts. Part I (Chapters 1–3) focuses on the history, concepts, and fundamentals of phase-based signal processing in speech communication. Part II (Chapters 4–6) is focused on the applications of phase-based processing for speech signal processing, while Chapter 7 summarizes the conclusions and provides some future research directions that phase-aware signal processing will follow. The book is completed by an appendix, which explains the MATLAB® toolbox called the *PhaseLab Toolbox*. In the following, the content of each individual chapter is briefly summarized.

Chapter 1 provides a full literature review on the historical viewpoints on the phase spectrum and the resulting beliefs on its importance/unimportance. Throughout the historical review we highlight the significance of phase information in human perception. The chapter addresses the fundamental questions, namely, is the spectral phase unimportant or important? If it is important, to what degree does the modification of the phase affect the performance of an algorithm? In addition, it is of interest to what extent the human auditory system is sensitive to distortions in the spectral phase. In order to address these questions, we will consider several key experiments through which we will examine the key statements made by researchers in favor or against the importance of the phase spectrum.

Chapter 2 presents a review of the fundamentals of phase-aware signal processing. We present the tools that are essential for understanding the key concepts and methods that are presented in the later chapters. These tools and fundamental concepts are helpful in developing novel phase-aware signal processing ideas, which require processing of the phase spectrum of a signal. In order to explain the main difficulties in phase process-ing which arise from the wrapping issue, the existing phase unwrapping methods will be discussed and compared by a comparative study. We will also highlight the useful representations of the phase spectrum that, in contrast to the conventional STFT phase spectrum, retain certain useful structures. Through some experiments, we will show how such phase representations are applicable as an effective tool, in order to access useful information about the spectral phase in phase-aware speech signal processing.

Chapter 3 is dedicated to the fundamentals of phase estimation. The chapter deals with the question of whether phase estimation is possible. As several key examples, we present the fundamental problem of estimating the phase of a harmonic in the pres-ence of adjacent harmonics and background noise, as well as the phase estimation error variance. A detailed analysis of the impact of the chosen window function on the phase estimation performance is presented. Later, the chapter provides several recommenda-tions of available methods for extracting the phase spectrum from a noisy observation. We will present an overview of previous and recent phase estimation methods proposed in the literature. The pros and cons of each category are considered. Finally, a compara-tive study is provided reporting the performance of different phase estimators in terms of their accuracy in estimating the clean speech phase spectrum.

The second part of the book focuses on presenting selected topics as representative examples in speech processing applications where phase-based signal processing has been successfully applied. The applications considered are single-channel speech enhancement (Chapter 4) and single-channel source separation (Chapter 5). In both chapters, conventional methods, i.e., phase-unaware solutions together with their justification, are presented first. The recent phase-aware signal processing solutions developed for the specific application are then explained. The importance of phase in these applications is emphasized using proof-of-concept experiments. Throughout the proof-of-concept experiments the improvements achievable in speech enhance-ment/separation by a phase-only modification approach or when combined with an enhanced spectral amplitude-only scheme will be quantified. A comparative study will be conducted to eventually quantify how much phase information could push the limits of the conventional phase-unaware solutions.

Chapter 6 addresses the topic of evaluating the performance of algorithms that modify phase as well as the spectral amplitude. The conventional instrumental metrics, often used to quantify the speech quality of a speech communication system, rely on spectral amplitude-only information. In fact, the reliability of these conventional met-rics when applied for phase-modified enhanced signals is questionable. The reliability of amplitude-only derived quality measures, therefore, is limited and conclusions on quality estimation for algorithms that modify the spectral phase only have to be drawn with caution. Chapter 6 addresses the following two questions: (i) are the existing instrumental measures reliable when applied for phase-aware performance evaluation? and (ii) do phase-aware measures possibly result in a more reliable correlation when predicting the subjective listening results? Several phase-aware instrumental metrics are considered as new candidates. Subjective listening results are reported for quality and intelligibility evaluation of phase-aware speech enhancement. A correlation analysis between the instrumental and subjective results clarifies which instrumental measures under which circumstances are reliable predictors of perceived speech quality or intelligibility.

Chapter 7 presents some concluding remarks and some possible future directions for research on phase processing.

The main body of this book consists of examples and figures included to help the reader to understand the theory and see how it is useful in practice. The book is supplemented with a toolbox in MATLAB® called the *PhaseLab Toolbox*. The toolbox comprises the implementations of the major phase-aware signal processing methods described in the book. In the appendix, we present a detailed description of the contents of the *PhaseLab Toolbox*. It is the strong belief of the authors that access to MATLAB® code and standard corpora is both necessary and crucial to bring an insightful understanding regarding the potential of the emerging field of phase-aware signal processing.

1.6 Experiments

In the following experiments, we examine the dispute about the *phase-deaf belief* and further investigate how Wang and Lim, Vary, and Paliwal came up with their observations and key statements about phase unimportance or importance as explained earlier in this chapter. The following experiments will be presented:

1) Phase unimportance in speech enhancement (Wang and Lim 1982).
2) Effects of phase modification (Vary 1985).
3) Mismatched window (Paliwal *et al.* 2011).
4) Phase spectrum compensation (Wojcicki *et al.* 2008).

These experiments will clarify when and how much phase information is important when dealing with an AMS framework for speech signals (see Figure 1.3).

1.6.1 Experiment 1.1: Phase Unimportance in Speech Enhancement

Figure 1.5 shows the block diagram for the experiment performed by Wang and Lim (1982), who focused on studying the importance of Fourier spectral phase in speech enhancement. In their experiment, they considered a test stimulus composed of amplitudes and phases contaminated at different SNRs using two noise sources, one for each channel, denoted by $d_1(n)$ and $d_2(n)$. Through listening experiments, the test stimulus was compared to a reference signal in order to determine an equivalent SNR. The sampling frequency was set to 10 kHz and the segment length was selected among 64, 512,

Figure 1.5 Block diagram for Wang and Lim's experiment (Wang and Lim 1982), where stimuli of phase-modified speech are constructed in the framework of analysis–modification–synthesis.

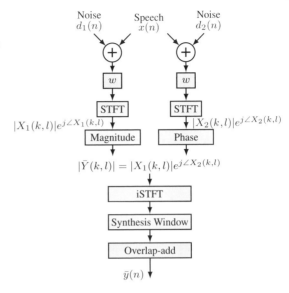

and 4096 samples in order to explore the impact of short, medium, and long window lengths. No zero padding was applied and a 50% overlap was used to reconstruct the time domain signal.

The STFT of the discrete-time signal denoted by $X(k, l)$ is given by

$$X(k, l) = \sum_{n=-\infty}^{\infty} x(n)w(n - lS)e^{-j\frac{2\pi kn}{N_{\mathrm{DFT}}}}, \tag{1.1}$$

with n and k referring to the discrete-time and frequency indices, S as the frame shift, l as the frame index, N_{DFT} as the length of the window in samples, and $w(n)$ as the prototype window function. The STFT coefficient $X(k, l)$ is complex and contains amplitude and phase spectra given by:

$$X(k, l) = |X(k, l)|e^{j\angle X(k, l)}, \tag{1.2}$$

where $|X(k, l)|$ and $\angle X(k, l)$ are defined as the spectral magnitude and phase, respectively.

To investigate the effects of phase modification, the phase-modified stimuli are generated as follows. Let $X_1(k, l)$ and $X_2(k, l)$ represent the first and the second paths' spectral amplitudes with $\angle X_1(k, l)$ and $\angle X_2(k, l)$ their corresponding spectral phases. By mixing the clean speech signal $x(n)$ with each noise signal, separately $d_1(n)$ and $d_2(n)$, we will obtain two noisy speech realizations at different SNRs, resulting in $|X_1(k, l)|e^{j\angle X_1(k, l)}$ and $|X_2(k, l)|e^{j\angle X_2(k, l)}$. The phase-modified complex spectrum stimuli $\overline{Y}(k, l)$, as shown in Figure 1.5, are a combination of the two paths:

$$\overline{Y}(k, l) = |X_1(k, l)|e^{j\angle X_2(k, l)}. \tag{1.3}$$

For synthesis, the least-squares overlap-add procedure (Quatieri 2001) is applied on the inverse STFT of the modified speech. The modified speech signal $\overline{y}(n)$ is reconstructed by applying inverse STFT on $\overline{Y}(k, l)$ and overlap-add procedure.

Table 1.1 Results for the Wang and Lim experiment in terms of SNR amplitude (SNR_a) versus SNR phase (SNR_p) showing the equivalent SNR (Wang and Lim 1982). The results are shown for a window length of 512 samples.

	SNR_p				
	−25	−5	5	15	25
SNR_a					
−25	−25	−25	−25	−25	−25
−5	—	−5	−3.9	−3.9	−4.0
5	—	—	5	4.9	6.0
15	—	—	—	15	16
25	—	—	—	—	—

They defined the equivalent SNR at which the reconstructed speech signal, either the modified speech or noisy unprocessed, were selected 50 percent of the time. The equivalent SNR calculated in the experiments by Wang and Lim quantifies the relative importance of phase and magnitude in speech enhancement. The test materials and the experimental procedure in this experiment are illustrated in Figure 1.5.

Table 1.1 shows the results reported by Wang and Lim. For window length $N_{DFT} = 4096$ and low SNR_a, an increase of SNR_p by 30 dB results in increasing the equivalent SNR by 11 dB. In contrast, for a short window length $N_{DFT} = 512$ and $SNR_a = −5$ dB, an increase of SNR_p by 30 dB results in an improvement of only 1 dB in the equivalent SNR. This coincides with the earlier reports by Oppenheim and Lim (1981). From this experiment, Wang and Lim concluded the following statements (Wang and Lim 1982):

- A less accurate phase estimate than the degraded phase can lead to a noticeable decrease in the equivalent SNR.
- A more accurate phase estimate than the degraded phase will not significantly improve the equivalent SNR.
- If a significantly different approach is used to exploit the phase information such as using the phase estimate to further improve the magnitude estimate, then a more accurate estimation of phase may be important.
- The phase impact becomes more important when the window length is larger. This is in agreement with the phase variance Cramer–Rao lower bound known from estimation theory, which decreases with larger data lengths.[8]

1.6.2 Experiment 1.2: Effects of Phase Modification

In 1985, Vary conducted an experiment to demonstrate the effects of phase modification (Vary 1985). Let us consider again the AMS framework shown in Figure 1.3. Figure 1.6 shows the block diagram for the test setup where a time domain clean speech signal $x(n)$ is passed through windowing and Fourier transformation to obtain its Fourier magnitude and phase spectra. The magnitude information is not modified, whereas the

---

8 For further details on this we refer to Kay (1993), or the discussion on phase estimation fundamentals given in Chapter 3.

Figure 1.6 Block diagram for Vary's experiment to study the effects of phase modification (Vary 1985).

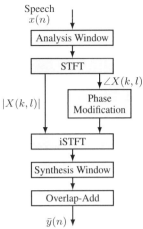

phase path is corrupted by additive noise. The inverse Fourier transform, followed by the overlap-and-add procedure, is used to reconstruct the time domain signal as explained in (1.4).

To study the effect of phase modification on the perceived quality of reconstructed speech, Vary undertook several experiments. First, by setting the spectral phase to zero, he observed that the reconstructed signal sounds *monotonous/voiced-only*. For the other extreme case, when completely randomizing the spectral phase, Vary reported that the sound was perceived as *unvoiced-excited-only*. As intermediate scenarios, Vary replaced the phase spectrum with a combination of the clean phase and the random phase. The degree of randomness was controlled by setting the variance of the noise added to the phase. Here, we consider $\frac{\pi}{2}, \frac{\pi}{4}, \frac{\pi}{6}$, and $\frac{\pi}{8}$ as the discrete candidates for the randomized phase values. The threshold of perception of the phase distortion was used by Vary to derive the corresponding SNR condition. It was demonstrated that an instantaneous SNR of 6 dB is the critical threshold.[9]

Vary defined phase deviation ϕ_{dev} as the difference between the noisy phase and clean phase, given by:

$$\phi_{\text{dev}}(k, l) = \phi_y(k, l) - \phi_x(k, l). \tag{1.4}$$

Figure 1.7 shows a vector diagram that demonstrates how phase deviation for frequency k is defined.

Vary in his derivations made two assumptions: (i) noise follows a Gaussian distribution, and (ii) he considered voiced segments only and assumed that the underlying harmonics are separated enough by the spectral analyzer. Under these assumptions, the maximum phase deviation was derived as:

$$\hat{\phi}_{\text{dev,max}} = \arcsin\left(\sqrt{\frac{\pi}{2} \frac{|D(k, l)|}{|X(k, l)|}}\right), \tag{1.5}$$

where k is the frequency index, while $D(k, l)$ and $X(k, l)$ refer to the noise and speech contributions for the kth frequency and lth frame, hence $\frac{|X(k,l)|^2}{|D(k,l)|^2}$ denotes the instantaneous SNR measured at local frequency k and frame index l.

---

9 Vary (1985) defined the local (in-band) SNR as $\frac{|X(k,l)|^2}{|D(k,l)|^2}$.

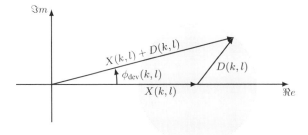

Figure 1.7 Vector diagram inspired by Vary (1985) showing the phase deviation $\phi_{\text{dev}}(k, l)$ resulting from the added noise to speech at frequency subband k and frame l.

For a local SNR of 6 dB, Vary derived the maximum phase deviation, $\hat{\phi}_{\text{dev,max}} = 0.679$ radians, corresponding roughly to the threshold of perception in phase distortion. From the listening experiments in Vary (1985), he concluded that the smallest detectable phase difference is between $\pi/8$ and $\pi/4$. Figure 1.8 shows the upper bound of the phase difference of the noisy and clean signals versus the input signal-to-noise ratio measured in dB. It is concluded that for a spectral SNR of larger than 6 dB a maximum phase deviation lower than $\pi/4$ is achievable. Vary concluded that when the magnitude information is sufficiently accurately estimated, no speech degradation due to phase distortion becomes perceptible as long as the local SNR is at least about 6 dB.

From this experiment, Vary made the following conclusions (Vary 1985):

- The difference only becomes perceptible when the phase deviation approaches $\frac{\pi}{4}$.
- For signal components with instantaneous SNRs above 6 dB, the noisy phase suffices and is a decent estimate for the desired clean phase.

The implementation is provided in the file *Exp1_2.m*, available in the *PhaseLab Toolbox* (see the appendix). It produces the results of Vary's experiment and can be used to justify the threshold of perception as maximum phase deviation equal to $\frac{\pi}{4}$.

1.6.3 Experiment 1.3: Mismatched Window

Paliwal *et al.* (2011) conducted an experiment that disputed the earlier observations reported by Wang and Lim. Figure 1.9 shows the block diagram used in their experimental setup. The clean speech signal enters two parallel paths which are different

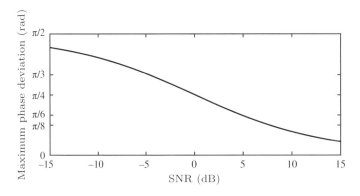

Figure 1.8 Phase deviation upper bound versus the spectral SNR in Vary's experiment given in (1.5).

Figure 1.9 Block diagram to construct stimuli of phase-modified speech in analysis–modification–synthesis (Paliwal *et al.* 2011).

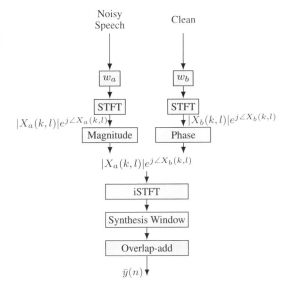

in terms of the window function used, distinguished by w_a and w_b. Through these two window functions and paths, Paliwal studied the impact of the dynamic range of the selected window on the accuracy of the phase representation. This effect was also quantified in terms of the resulting perceived quality in the reconstructed speech signal in the AMS framework. In order to extend their experiment to different noisy conditions controlled by SNR, in the first path they included a noisy speech using white noise. The modified signal spectra of the path using windows w_a and w_b are denoted $X_a(k,l) = |X_a(k,l)|e^{j\angle X_a(k,l)}$ and $X_b(k,l) = |X_b(k,l)|e^{j\angle X_b(k,l)}$, respectively. The length of both windows was set to 32 ms and SNRs of 0 and 10 dB were studied. The following four treatments were considered:

- **Wang-O:** The overlap was set to 50%, and both window functions were set to the Hann window. No zero padding was used.
- **Matched-O:** An overlap of 87.5% was chosen with both windows set to Hamming ($w_a = w_b$), hence matched. No zero padding was used.
- **Mismatched-O:** Overlap of 87.5%. The first path window w_a was set to Hamming while the other one w_b was set to the Chebyshev window, hence the two paths are mismatched in terms of the dynamic range. Zero padding was applied.
- **Mismatched-N:** Similar to Mismatched-O, only the oracle phase is replaced with the noisy phase at the reconstruction stage.

Table 1.2 shows the results as the *mean opinion score* (MOS) for each of the treatment types described above. The results are shown for (first row) SNR = 0 dB and (second row) SNR = 10 dB. The following conclusions were made (Paliwal *et al.* 2011):

- If overlap is increased and zero padding is applied, Paliwal reported significant improved perceived speech quality.
- Applying mismatched windows results in improvement in the MOS instrumentally predicted quality measured in PESQ (Rix *et al.* 2001).

Table 1.2 Subjective speech quality for different SNRs and different treatment types with SNR of 0 and 10 dB (Paliwal *et al.* 2011).

Treatment type	Clean	Noisy	Matched-O	Mismatched-N	Mismatched-O	PSC
SNR = 0 dB	1.00	0.08 ± 0.12	0.25 ± 0.12	0.28 ± 0.12	0.75 ± 0.02	0.55 ± 0.05
SNR = 10 dB	1.00	0.08 ± 0.08	0.28 ± 0.12	0.26 ± 0.1	0.74 ± 0.02	0.6 ± 0.1

- For the mismatched-O stimuli, there exists a trade-off between the reservedness of the fine spectral detail and the amount of noise suppression. This trade-off is controlled through the dynamic range of the window w_b.

Following these observations and experiments, Paliwal *et al.* made the following key statements:

- Research into better phase spectrum estimation algorithms, while a challenging task, could be worthwhile.
- Accurate phase spectrum estimates have the potential to significantly improve the performance of the existing magnitude-spectrum-based methods.

The MATLAB® file *Exp1_3.m* available in the *PhaseLab Toolbox* reproduces some of the results reported by Paliwal, where an improved PESQ score was observed by choosing a mismatched window at the analysis and synthesis stages for noise-corrupted speech with white noise.

1.6.4 Experiment 1.4: Phase Spectrum Compensation

It is known that the DFT spectrum of a real signal (e.g., speech) is conjugate symmetric. Taking this fact into account, Stark *et al.* (2008) proposed a speech enhancement method relying on controlled violation of this conjugate symmetricity enforced on the noisy spectrum. To this end, the phase spectrum compensation function denoted by $\Lambda(k, l)$ was used to alter the angular relationship between the DFT spectra to reinforce the conjugate symmetricity.

Figure 1.10 visualizes the procedure carried out by the phase spectrum compensation (PSC) method in terms of a vector diagram for a single conjugate pair. The results are shown for two extreme scenarios: (a) low and (b) high SNR, whereby, depending on the magnitude of $Y(k, l)$ and $\Lambda(k, l)$, the strength of compensation varies. The phase-compensated noisy STFT before signal reconstruction stage is then given by:

$$Y_\Lambda(k, l) = Y(k, l) + \Lambda(k, l), \tag{1.6}$$

where $\Lambda(k, l)$ was defined as (Stark *et al.* 2008):

$$\Lambda(k, l) = \lambda \Upsilon(k, l) |\hat{D}(k, l)|, \tag{1.7}$$

where $|\hat{D}(k, l)|$ is the estimated noise magnitude spectrum and $\Upsilon(k, l)$ is the anti-symmetry function defined as:

$$\Upsilon(k, l) = \begin{cases} 1 & \text{if} \quad 0 < k/N_{\text{DFT}} < 0.5 \\ -1 & \text{if} \quad 0.5 < k/N_{\text{DFT}} < 1 \\ 0 & \text{otherwise.} \end{cases} \tag{1.8}$$

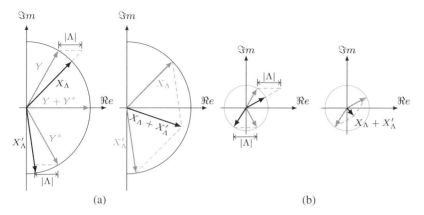

(a)   (b)

Figure 1.10 Vector diagrams showing phase spectrum compensation (PSC) inspired by Stark *et al.* (2008), where modification of the noisy STFT is shown for conjugate pair signal-to-noise ratio scenarios: (a) large ($|Y| > |\Lambda|$), (b) low ($|Y| < |\Lambda|$).

The compensated phase spectrum is finally used together with the noisy spectral amplitude in order to produce an enhanced complex domain spectrum:

$$\hat{X}_\Lambda(k, l) = |Y(k, l)|e^{j\angle Y_\Lambda(k,l)}. \tag{1.9}$$

Using the inverse DFT followed by an overlap-add procedure and taking the real part, a time domain enhanced speech signal is produced. The block diagram for the PSC method is similar to Figure 1.6; the phase modification is carried out by adding the asymmetric function $\Lambda(k, l)$ given by (1.6).

The noisy speech is produced by adding white noise to a male speech file at 10 dB signal-to-noise ratio. In this experiment, 87.5% overlap was used with a sampling frequency of 16 kHz. The parameter $\lambda = 3.74$ was used following the recommendation in Stark *et al.* (2008) and Wojcicki *et al.* (2008). The results are shown in Figure 1.11 as spectrograms for the clean (top), noisy (middle), and enhanced (bottom) signals. The improvement in speech quality measured in PESQ and SNR are reported at the top of each panel. PSC is well suited for mid- to high-SNR scenarios and shows the importance

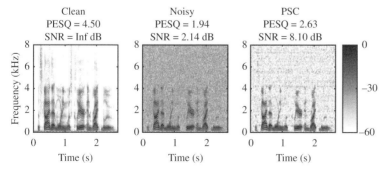

Figure 1.11 Speech enhancement using phase spectrum compensation (PSC; Stark *et al.* 2008). Spectrograms in dB are shown for (left) clean, (middle) noisy, (right) enhanced speech signals. PESQ and output SNR scores are shown at the top of each panel.

of phase spectral information for speech enhancement. For the experiment, we used the software from *psc.m* developed in Stark *et al.* (2008).

1.7 Summary

In this chapter, we provided a full literature review on the rise, decline, and renaissance of phase processing for speech applications, consisting of a chronologically ordered discussion of phase-related reports, experiments, and attempts made by researchers throughout the last century. The two viewpoints on the unimportance or importance of phase were discussed, with several experiments and simulations provided to illustrate these beliefs. The outcomes demonstrated the significance and potential of the phase spectrum in speech processing. The chapter concluded with statements answering the question of when and how much phase information may have an impact on the perceived quality of a processed speech signal.

References

P. Aarabi and G. Shi, Phase-based dual-microphone robust speech enhancement, *IEEE Transactions on Systems, Man, and Cybernetics Part B: Cybernetics*, vol. 34, no. 4, pp. 1763–1773, 2004.

P. Aarabi, G. Shi, M. M. Shanechi, and A. Seyed Rabi, *Phase-Based Speech Processing*, World Scientific Publishing, 2006.

Y. Agiomyrgiannakis and Y. Stylianou, Wrapped Gaussian mixture models for modeling and high-rate quantization of phase data of speech, *IEEE Transactions on Audio, Speech, and Language Processing*, vol. 17, no. 4, pp. 775–786, 2009.

Y. Agiomyrgiannakis, *VOCAINE: The Vocoder and Applications in Speech Synthesis*, Proceedings of the IEEE International Conference on Acoustics, Speech and Signal Processing (ICASSP), pp. 4230–4234, 2015.

L. D. Alsteris and K. K. Paliwal, *ASR on Speech Reconstructed from Short-Time Fourier Phase Spectra*, Proceedings of the International Conference on Spoken Language Processing (INTERSPEECH), pp. 565–568, 2004.

L. D. Alsteris and K. K. Paliwal, Further intelligibility results from human listening tests using the short-time phase spectrum, *Speech Communication*, vol. 48, no. 6, pp. 727–736, 2006.

L. D. Alsteris and K. K. Paliwal, Short-time phase spectrum in speech processing: A review and some experimental results, *Signal Processing*, vol. 17, no. 3, pp. 578–616, 2007.

F. A. Bilsen, On the influence of the number and phase of harmonics on the perceptibility of the pitch of complex signals, *Acta Acustica*, vol. 28, no. 1, pp. 50–65, 1973.

M. Beutnagel, A. Conkie, J. Schroeter, Y. Stylianou, and A. Syrdal, *The AT&T Next-Gen TTS System*, Proc. Joint meeting of ASA, EAA, and DAGA, pp. 18–24, 1999.

M. S. Brandstein, P. A. Monta, J. C. Hardwick, and J. S. Lim, *A Real-Time Implementation of the Improved MBE Speech Coder*, Proceedings of the IEEE International Conference on Acoustics, Speech and Signal Processing (ICASSP), vol. 1, pp. 5–8, 1990.

R. Carlson, B. Granstrom, D. Klatt, *Vowel Perception: The Relative Perceptual Salience of Selected Acoustic Manipulations*, Quarterly Progress Status Report, Speech Transmission Lab., Royal Institute of Technology, vol. 3, pp. 73–83, 1979.

S. Davis and P. Mermelstein, Comparison of parametric representations for monosyllabic word recognition in continuously spoken sentences, *IEEE Transactions on Acoustics, Speech, and Signal Processing*, vol. 28, no. 4, pp. 357–366, 1980.

M. Dolson, The phase vocoder: A tutorial, *Computer Music Journal*, vol. 10, no. 4, pp. 14–27, 1986.

X. Dong, M. F. Bocko, and Z. Ignjatovic, *Data Hiding via Phase Manipulation of Audio Signals*, Proceedings of the IEEE International Conference on Acoustics, Speech and Signal Processing (ICASSP), pp. 377–380, 2004.

C. Espy and J. S. Lim, *Effects of Noise on Signal Reconstruction from Fourier Transform Phase*, Proceedings of the IEEE International Conference on Acoustics, Speech and Signal Processing (ICASSP), pp. 1833–1836, 1982.

B. Geiser and P. Vary, Artificial bandwidth extension of wideband speech by pitch-scaling of higher frequencies, *Workshop Audiosignal- und Sprachverarbeitung (WASP)*, vol. 220, pp. 2892–2901, 2013.

T. Gerkmann, M. B. Kwayczyk, and J. Le Roux, Phase processing for single-channel speech enhancement: History and recent advances, *IEEE Signal Processing Magazine*, vol. 32, no. 2, pp. 55–66, 2015.

P. Golik, Z. Tüske, R. Schlüter, and H. Ney, *Convolutional Neural Networks for Acoustic Modeling of Raw Time Signal in LVCSR*, Proceedings of the International Conference on Spoken Language Processing (INTERSPEECH), pp. 26–30, 2015.

D. Griffin and J. Lim, Signal estimation from modified short-time Fourier transform, *IEEE Transactions on Acoustics, Speech, and Signal Processing*, vol. 32, no. 2, pp. 236–243, 1984.

S. Guangji, M. M. Shanechi, and P. Aarabi, On the importance of phase in human speech recognition, *IEEE Transactions on Audio, Speech, and Language Processing*, vol. 14, no. 5, pp. 1867–1874, 2006.

D. Guozhen and R. A. Morris, The importance of phase in the spectra of digital type, *Electronic Publishing*, vol. 2 no. 1 1989.

M. Hayes, J. Lim, and A. V. Oppenheim, Signal reconstruction from phase or magnitude, *IEEE Transactions on Acoustics, Speech, and Signal Processing*, vol. 28, no. 6, pp. 672–680, 1980.

R. M. Hegde, H. A. Murthy, and G. V. R. Rao, *Application of the Modified Group Delay Function to Speaker Identification and Discrimination*, Proceedings of the IEEE International Conference on Acoustics, Speech and Signal Processing (ICASSP), pp. 517–520, 2004.

F. Heese, B. Geiser, and P. Vary, *Intelligibility Assessment of a System for Artifical Bandwidth Extension of Telephone Speech*, Proceedings of the German Annual Conference on Acoustics (DAGA), pp. 905–906, 2010.

H. von Helmholtz, *On the Sensations of Tone as a Physiological Basis for the Theory of Music*, Longmans, Green and Co., 1912.

I. Hernaez, I. Saratxaga, J. Ye, J. Sanchez, D. Erro, and E. Navas, Speech watermarking based on coding of the harmonic phase, *Advances in Speech and Language Technologies for*

Iberian Languages, Lecture Notes in Computer Science, vol. 8854, pp. 259–268, 2014.

K. Hofbauer, G. Kubin, and W. B. Kleijn, Speech watermarking for analog flat-fading bandpass channels, *IEEE Transactions on Audio, Speech, and Language Processing*, vol. 17, no. 8, pp. 1624–1637, 2009.

M. Honda, *Speech Analysis-Synthesis Using Phase-Equalized Excitation*, Technical Report of IEICE, pp. 1–8, 1989.

X. Huang, A. Acero, and H.-W. Hon, *Spoken Language Processing: A Guide to Theory, Algorithm, and System Development*, Prentice Hall, 2001.

B. Iser, G. Schmidt, and W. Minker, *Bandwidth Extension of Speech Signals*, Lecture Notes in Electrical Engineering, Springer, 2008.

P. Jax, *Enhancement of Bandlimited Speech Signals: Algorithms and Theoretical Bounds*, PhD Thesis, IND, RWTH Aachen, 2002.

S. M. Kay, *Fundamentals of Statistical Signal Processing: Estimation theory*, Prentice Hall, 1993.

M. Kazama, S. Gotoh, M. Tohyama, and T. Houtgast, On the significance of phase in the short term Fourier spectrum for speech intelligibility, *The Journal of the Acoustical Society of America*, vol. 127, no. 3, pp. 1432–1439, 2010.

D. S. Kim, *Perceptual Phase Redundancy in Speech*, Proceedings of the IEEE International Conference on Acoustics, Speech and Signal Processing (ICASSP), vol. 3, pp. 1383–1386, 2000.

D. S. Kim, Perceptual phase quantization of speech, *IEEE Transactions on Speech and Audio Processing*, vol. 11, no. 4, pp. 355–364, 2003.

W. B. Kleijn and K. K. Paliwal, *Speech Coding and Synthesis*, Elsevier Science, 1995.

T. Kleinschmidt, S. Sridharan, and M. Mason, The use of phase in complex spectrum subtraction for robust speech recognition, *Computer Speech & Language*, vol. 25, no. 3, pp. 585–600, 2011.

G. Kubin, B. S. Atal, and W. B. Kleijn, *Performance of Noise Excitation for Unvoiced Speech*, Proceedings of IEEE International Workshop on Speech Coding for Telecommunications, pp. 35–36, 1993.

S. S. Kuo, J. D. Johnston, T. William, and S. R. Quackenbush, *Covert Audio Watermarking using Perceptually Tuned Signal Independent Multiband Phase Modulation*, Proceedings of the IEEE International Conference on Acoustics, Speech and Signal Processing (ICASSP), pp. 1753–1756, 2002.

M. V. Laitinen, S. Disch, and V. Pulkki, Sensitivity of human hearing to changes in phase spectrum, *The Journal of the Acoustical Society of America*, vol. 61, no. 11, pp. 860–877, 2013.

P. L. de Leon, I. Hernaez, I. Saratxaga, M. Pucher, and J. Yamagishi, *Detection of Synthetic Speech for the Problem of Imposture*, Proceedings of the IEEE International Conference on Acoustics, Speech and Signal Processing (ICASSP), pp. 4844–4847, 2011.

P. Y. Liew and M. A. Armand, Inaudible watermarking via phase manipulation of random frequencies, *Multimedia Tools and Applications*, vol. 35, no. 3, pp. 357–377, 2007.

S. P. Lipshitz, M. Pocock, and J. Vanderkooy, On the audibility of midrange phase distortion in audio systems, *Audio Engineering Society*, vol. 30, no. 9, pp. 580–595, 1982.

L. Liu, H. Jialong, and P. Günther, Effects of phase on the perception of intervocalic stop consonants, *Speech Communication*, vol. 22, no. 4, pp. 403–417, 1997.

E. Loweimi, J. Barker, and T. Hain, *Source-Filter Separation of Speech Signal in the Phase Domain*, Proceedings of the International Conference on Spoken Language Processing (INTERSPEECH), pp. 598–602, 2015.

E. Loweimi and S. M. Ahadi, *A New Group Delay-Based Feature for Robust Speech Recognition*, Proceedings of the IEEE International Conference on Multimedia and Expo (ICME), pp. 1–5, 2011.

E. Loweimi, S. M. Ahadi, and T. Drugman, *A New Phase-Based Feature Representation for Robust Speech Recognition*, Proceedings of the IEEE International Conference on Acoustics, Speech and Signal Processing (ICASSP), pp. 7155–7159, 2013.

R. Maia, Y. Stylianou, Iterative Estimation of Phase Using Complex Cepstrum Representation, *Proceedings of the IEEE International Conference on Acoustics, Speech and Signal Processing (ICASSP)*, 2016.

R. McAulay and T. F. Quatieri, Speech analysis/synthesis based on a sinusoidal representation, *IEEE Transactions on Acoustics, Speech, and Signal Processing*, vol. 34, no. 4, pp. 744–754, 1986.

I. McCowan, D. Dean, M. McLaren, R. Vogt, and S. Sridharan, The delta-phase spectrum with application to voice activity detection and speaker recognition, *IEEE Transactions on Audio, Speech, and Language Processing*, vol. 19, no. 7, pp. 2026–2038, 2011.

R. Maia, M. Akamine, and M. J. F. Gales, *Complex Cepstrum as Phase Information in Statistical Parametric Speech Synthesis*, Proceedings of the IEEE International Conference on Acoustics, Speech and Signal Processing (ICASSP), pp. 4581–4584, 2012.

R. Maia and Y. Stylianou, *Complex Cepstrum Factorization for Statistical Parametric Synthesis*, Proceedings of the IEEE International Conference on Acoustics, Speech and Signal Processing (ICASSP), pp. 3839–3843, 2014.

A. Maly and P. Mowlaee, *On the Importance of Harmonic Phase Modification for Improved Speech Signal Reconstruction*, Proceedings of the IEEE International Conference on Acoustics, Speech and Signal Processing (ICASSP), 2016.

F. Mayer and P. Mowlaee, *Improved Phase Reconstruction in Single-Channel Speech Separation*, Proceedings of the International Conference on Spoken Language Processing (INTERSPEECH), 2015.

P. Mowlaee and J. Kulmer, Phase estimation in single-channel speech enhancement: Limits-potential, *IEEE/ACM Transactions on Audio, Speech, and Language Processing*, vol. 23, no. 8, pp. 1283–1294, 2015a.

P. Mowlaee and J. Kulmer, Harmonic phase estimation in single-channel speech enhancement using phase decomposition and SNR information, *IEEE/ACM Transactions on Audio, Speech, and Language Processing*, vol. 23, no. 9, pp. 1521–1532, 2015b.

K. S. R. Murty and B. Yegnanarayana, Combining evidence from residual phase and MFCC features for speaker recognition, *IEEE Signal Processing Letters*, vol. 13, no. 1, pp. 52–55, 2006.

S. Nakagawa, K. Asakawa, and L. Wang, *Speaker Recognition by Combining MFCC and Phase Information*, Proceedings of the International Conference on Spoken Language Processing (INTERSPEECH), pp. 2005–2008, 2006.

S. Nakagawa, L. Wang, and S. Ohtsuka, Speaker identification and verification by combining MFCC and phase information, *IEEE Transactions on Audio, Speech, and Language Processing*, vol. 20, no. 4, pp. 1085–1095, 2012.

X. Ni and X. Huo, Statistical interpretation of the importance of phase information in signal and image reconstruction, *Statistics & Probability Letters*, vol. 77, no. 4, pp. 447–454, 2007.

M. Nilsson and W. B. Kleijn, *Avoiding Over-Estimation in Bandwidth Extension of Telephony Speech*, Proceedings of the IEEE International Conference on Acoustics, Speech and Signal Processing (ICASSP), pp. 869–872, 2001.

G. S. Ohm, Über die Definition des Tones, nebst daran geknüfter Theorie der Sirene und ähnlicher tonbildender Vorichtungen, *Journal of Physical Chemistry*, vol. 59, pp. 513–565, 1843.

A. V. Oppenheim, J. S. Lim, G. Kopec, and S. C. Pohlig, *Phase in Speech and Pictures*, Proceedings of the IEEE International Conference on Acoustics, Speech and Signal Processing (ICASSP), vol. 4, pp. 632–637, 1979.

A. V. Oppenheim and J. S. Lim, The importance of phase in signals, *Proceedings of the IEEE*, vol. 69, no. 5, pp. 529–541, 1981.

A. V. Oppenheim, J. S. Lim, and S. R. Curtis, Signal synthesis and reconstruction from partial Fourier-domain information, *The Journal of the Acoustical Society of America*, vol. 73, no. 11, pp. 1413–1420, 1983.

K. K. Paliwal and L. D. Alsteris, *Usefulness of Phase Spectrum in Human Speech Perception*, Proceedings of the European Conference on Speech Communication and Technology, pp. 2117–2120, 2003.

K. K. Paliwal and L. D. Alsteris, On the usefulness of STFT phase spectrum in human listening tests, *Speech Communication*, vol. 45, no. 2, pp. 153–170, 2005.

K. K. Paliwal, K. K. Wojcicki, and B. J. Shannon, The importance of phase in speech enhancement, *Speech Communication*, vol. 53, no. 4, pp. 465–494, 2011.

R. D. Patterson, A pulse ribbon model of monaural phase perception, *The Journal of the Acoustical Society of America*, vol. 82, no. 5, pp. 1560–1586, 1987.

R. D. Patterson, The sound of a sinusoid: Time-interval models, *The Journal of the Acoustical Society of America*, vol. 96, no. 3, pp. 1419–1428, 1994.

R. Peharz, G. Kapeller, P. Mowlaee, and F. Pernkopf, *Modeling Speech with Sum-Product Networks: Application to Bandwidth Extension*, Proceedings of the IEEE International Conference on Acoustics, Speech and Signal Processing (ICASSP), pp. 3699–3703, 2014.

J. W. Picone, Signal modeling techniques in speech recognition, *Proceedings of the IEEE*, vol. 81, no. 9, pp. 1215–1247, 1993.

R. Plomp and H. J. M. Steeneken, Effect of phase on the timbre of complex tones, *The Journal of the Acoustical Society of America*, vol. 46, pp. 409–421, 1969.

H. Pobloth and W. B. Kleijn, *On Phase Perception in Speech*, Proceedings of the IEEE International Conference on Acoustics, Speech and Signal Processing (ICASSP), pp. 29–32, 1999.

H. Pobloth and W. B. Kleijn, Squared error as a measure of perceived phase distortion, *The Journal of the Acoustical Society of America*, vol. 114, no 2, pp. 1081–1094, 2003.

H. Pobloth, *Perceptual and Squared Error Aspects in Speech and Audio Coding*, PhD Thesis, Royal Institute of Technology (KTH), 2004.

H. Pulakka, U. Remes, S. Yrttiaho, K. Palomaki, M. Kurimo, and P. Alku, Bandwidth extension of telephone speech to low frequencies using sinusoidal synthesis and a Gaussian mixture model, *IEEE Transactions on Audio, Speech, and Language Processing*, vol. 20, no. 8, pp. 2219–2231, 2012.

T. F. Quatieri, *Phase Estimation with Application to Speech Analysis-Synthesis*, Technical Report, Massachusetts Institute of Technology, Research Laboratory of Electronics, 1979.

T. F. Quatieri, *Discrete-Time Speech Signal Processing: Principles and Practice*, Prentice Hall, 2001.

T. F. Quatieri and A. V. Oppenheim, Iterative techniques for minimum phase signal reconstruction from phase or magnitude, *IEEE Transactions on Acoustics, Speech, and Signal Processing*, vol. 29, no. 6, pp. 1187–1193, 1981.

A. W. Rix, J. G. Beerends, M. P. Hollier, and A. P. Hekstra, *Perceptual Evaluation of Speech Quality (PESQ): A New Method for Speech Quality Assessment of Telephone Networks and Codecs*, Proceedings of the IEEE International Conference on Acoustics, Speech and Signal Processing (ICASSP), vol. 2, pp. 749–752, 2001.

J. Sanchez, I. Saratxaga, I. Hernaez, E. Navas, D. Erro, and T. Raitio, Toward a universal synthetic speech spoofing detection using phase information, *IEEE Transactions on Information Forensics and Security*, vol. 10, no. 4, pp. 810–820, 2015.

S. O. Sadjadi and J. H. L. Hansen, *Hilbert Envelope based Features for Robust Speaker Identification under Reverberant Mismatched Conditions*, Proceedings of the IEEE International Conference on Acoustics, Speech and Signal Processing (ICASSP), pp. 5448–5451, 2011.

I. Saratxaga, I. Hernaez, M. Pucher, and I. Sainz, *Perceptual Importance of the Phase Related Information in Speech*, Proceedings of the International Conference on Spoken Language Processing (INTERSPEECH), pp. 1448–1451, 2012.

R. Schlüter and H. Ney, *Using Phase Spectrum Information for Improved Speech Recognition Performance*, Proceedings of the IEEE International Conference on Acoustics, Speech and Signal Processing (ICASSP), pp. 133–136, 2001.

M. R. Schroeder, New results concerning monaural phase sensitivity, *The Journal of the Acoustical Society of America*, vol. 31, no. 11, 1959.

M. R. Schroeder, Models of hearing, *Proceedings of the IEEE*, vol. 63, no. 9, pp. 1332–1350, 1975.

M. R. Schroeder and H. W. Strube, Flat-spectrum speech, *The Journal of the Acoustical Society of America*, vol. 79, no. 5, pp. 1580–1583, 1986.

G. Shi, P. Aarabi, and J. Hui, Phase-based dual-microphone speech enhancement using a prior speech model, *IEEE Transactions on Audio, Speech, and Language Processing*, vol. 15, no. 1, pp. 109–118, 2007.

J. Skoglund, W. B. Kleijn, and P. Hedelin, *Audibility of Pitch-Synchronously Modulated Noise*, Proceedings of the IEEE Workshop on Speech Coding For Telecommunications Proceeding, pp. 51–52, 1997.

A. P. Stark and K. K. Paliwal, *Speech Analysis using Instantaneous Frequency Deviation*, Proceedings of the International Conference on Spoken Language Processing (INTERSPEECH), pp. 22–26, 2008.

A. P. Stark and K. K. Paliwal, *Group-Delay-Deviation based Spectral Analysis of Speech*, Proceedings of the International Conference on Spoken Language Processing (INTERSPEECH), pp. 1083–1086, 2009.

A. P. Stark, K. K. Wojcicki, J. G. Lyons, and K. K. Paliwal, *Noise Driven Short Time Phase Spectrum Compensation Procedure for Speech Enhancement*, Proceedings of the International Conference on Spoken Language Processing (INTERSPEECH), pp. 549–552, 2008.

Y. Stylianou, Applying the harmonic plus noise model in concatenative speech synthesis, *IEEE Transactions on Speech and Audio Processing*, vol. 9, no. 1, pp. 21–29, 2001.

X. Tian, S. Du, X. Xiao, H. Xu, E. S. Chng, and H. Li, *Detecting Synthetic Speech using Long Term Magnitude and Phase Information*, Proceedings of the IEEE China Summit and International Conference on Signal and Information Processing (ChinaSIP), pp. 611–615, 2015.

R. Steven Turner, The Ohm–Seebeck dispute, Hermann von Helmholtz, and the origins of physiological acoustics, *The British Journal for the History of Science*, vol. 10, pp 1–24, 1977.

Z. Tüske, P. Golik, R. Schlüter, and H. Ney, *Acoustic Modeling with Deep Neural Networks Using Raw Time Signal for LVCSR*, Proceedings of the International Conference on Spoken Language Processing (INTERSPEECH), pp. 890–894, 2014.

P. Vary, Noise suppression by spectral magnitude estimation mechanism and theoretical limits, *Signal Processing*, vol. 8, no. 4, pp. 387–400, 1985.

P. Vary and R. Martin, *Digital Speech Transmission: Enhancement, Coding And Error Concealment*, John Wiley & Sons, 2006.

D. Wang and J. Lim, The unimportance of phase in speech enhancement, *IEEE Transactions on Acoustics, Speech, and Signal Processing*, vol. 30, no. 4, pp. 679–681, 1982.

L. Wang, K. Minami, K. Yamamoto, and S. Nakagawa, *Speaker Identification by Combining MFCC and Phase Information in Noisy Environments*, Proceedings of the IEEE International Conference on Acoustics, Speech and Signal Processing (ICASSP), pp. 4502–4505, 2010.

M. R. Weiss, E. Aschkenasy, and T. W. Parsons, *Study and Development of the INTEL Technique for Improving Speech intelligibility*, Technical Report, Nicolet Scientific Corp., 1974.

K. K. Wojcicki, M. Milacic, A. Stark, J. G. Lyons, and K. K. Paliwal, Exploiting conjugate symmetry of the short-time Fourier spectrum for speech enhancement, *IEEE Signal Processing Letters*, vol. 15, pp. 461–464, 2008.

Z. Wu, N. Evans, T. Kinnunen, J. Yamagishi, F. Alegre, and H. Li, Spoofing and countermeasures for speaker verification: A survey, *Speech Communication*, vol. 66, pp. 130–153, 2015.

X. Xiao, X. Tian, S. Du, H. Xu, E. Chng, and H. Li, *Spoofing Speech Detection using High Dimensional Magnitude and Phase Features: the NTU Approach for ASVspoof 2015 challenge*, Proceedings of the International Conference on Spoken Language Processing (INTERSPEECH), pp. 2052–2056, 2015.

E. W. M. Yu and C. F. Chan, *Harmonic + Noise Coding based on the Characteristics of Human Phase Perception*, Proceedings of the International Symposium on Intelligent Multimedia, Video and Speech Processing, pp. 310–313, 2004.

H. Zen, T. Nose, J. Yamagishi, S. Sako, T. Masuko, A. Black, and K. Tokuda, *The HMM-Based Speech Synthesis System (HTS) Version 2.0*, Proc. International Speech Communication Association Speech Synthesis Workshop (SSW6), pp. 294–299, 2006.

H. Zwicker and E. Fastl, *Psychoacoustics: Facts and Models*, Springer Series in Information Sciences, Springer, 2007.

2

Fundamentals of Phase-Based Signal Processing

Pejman Mowlaee

Graz University of Technology, Graz, Austria

2.1   Chapter Organization

In this chapter, the objective is to provide a compilation of practical concepts and useful analysis tools for phase-based signal processing. The aim is to make spectral phase accessible for researchers working on speech signal processing. Further, this knowledge will be useful in understanding the phase-based signal processing ideas explained in the following chapters. In particular, knowledge from this chapter is required to fully understand the phase-aware processing solutions discussed for the applications in the second part of the book. The major problem in phase signal processing is the phase wrapping in the spectral Fourier analysis. Examples are given in order to visualize the phase wrapping problem and further to demonstrate how phase unwrapping works. To reveal this in more detail, several phase unwrapping methods will be discussed, with a comparative study demonstrating their effectiveness.

Due to the wrapping, the Fourier phase spectrum will not produce an accessible or useful presentation. As analysis tools, the chapter explains several useful representations to display the phase information in a signal. These tools include: *group delay* (GD), *instantaneous frequency* (IF), *phase variance*, harmonic unwrapped phase, *phase distortion*, *baseband phase difference* (BPD), and *relative phase shift* (RPS). Through computer simulations, the concepts and tools will be explained in detail. We will also present a comparative study between the phase-based features presented in this chapter.

2.2   STFT Phase: Background and Some Remarks

2.2.1   Short-Time Fourier Transform

The time domain representation of a signal often provides only raw information. In contrast, using a spectral transformation enables access to specific properties of the signal.

Single Channel Phase-Aware Signal Processing in Speech Communication: Theory and Practice, First Edition. Pejman Mowlaee, Josef Kulmer, Johannes Stahl, and Florian Mayer.

An appropriate representation of the underlying signal is a key factor in the analysis and design of a signal processing solution for a given application. Among the various choices for signal transformations, the STFT is a popular analysis tool due to its simplicity and efficient hardware implementation.

Let n be the discrete time index and $\omega = \frac{2\pi f}{f_s}$ the normalized angular frequency, with f the frequency index in Hertz, $f_s = 1/T_s$ the sampling frequency, and T_s the sampling period. The discrete-time signal $x(n)$ is segmented into short durations at each frame index l denoted by $x_l(n)$ of length N. The Fourier transform of the lth frame $x_l(n)$ is given by:

$$X(e^{j\omega}) = \sum_{n=-\infty}^{\infty} x_l(n)e^{-j\omega n}. \tag{2.1}$$

Following the sampling theorem, for the Fourier transform of $x_l(n)$ we have (Oppenheim and Schafer 1989):

$$x_l(n) = \frac{1}{2\pi} \int_{-\pi}^{\pi} X(e^{j\omega})e^{jn\omega} d\omega. \tag{2.2}$$

The discrete-time Fourier transform $X(e^{j\omega})$ is complex-valued, containing magnitude and phase parts, which can be represented as a phasor in a polar format; thus we have

$$X(e^{j\omega}) = |X(e^{j\omega})|e^{j\phi(e^{j\omega})}, \tag{2.3}$$

where $|X(e^{j\omega})|$ is the magnitude spectrum and $\phi(e^{j\omega}) = \arg[X(e^{j\omega})]$ is the continuous phase spectrum. The principal value of the phase spectrum lies in the range $[-\pi, \pi[$ and is denoted by $\mathrm{ARG}[X(e^{j\omega})]$. In practice, the DFT is used instead of the *discrete-time Fourier transform* (DFT). The DFT is defined as the sampled version of the DTFT, sampled at $\omega_k = \frac{2\pi k}{N_{\mathrm{DFT}}}$, with N_{DFT} defined as the number of DFT points. We define k as the discrete frequency index, with $k \in [0, N_{\mathrm{DFT}} - 1]$, that is given by sampling the continuous Fourier frequency which is also interpreted as the center frequency of the STFT bands.

2.2.2 Fourier Analysis of Speech: STFT Amplitude and Phase

In many speech processing applications, it is quite common to employ the Fourier transformation for feature extraction or for spectral analysis purposes. The Fourier transformation is applied to the segmented windowed speech as Fourier analysis requires stationarity of the underlying signal. The outcomes of such a short-time Fourier analysis are the instantaneous STFT magnitude and phase spectra.

To show the use of the STFT as a spectral analyzer to represent a given speech signal, we continue with an example. Figure 2.1 shows the time domain waveform of a clean speech signal $x(n)$ (left), magnitude spectrum $|X(e^{j\omega})|$ (middle), and phase spectrum $\phi(e^{j\omega}) = \mathrm{ARG}[X(e^{j\omega})]$ (right). The result is shown for a female utterance selected from the Texas Instruments and Massachusetts Institute of Technology (TIMIT) database (Garofolo *et al.* 1993). The utterance is "December and January are nice months to spend in Miami," composed of various phoneme classes including plosives [d, b, p], fricatives [s], nasals, and several vowels. The time domain waveform shown in the left panel presents the overall distribution of signal energy over time. The spectrogram magnitude shown in the middle panel provides information about the individual frequency

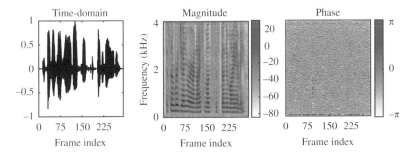

Figure 2.1 Time domain (left), magnitude spectrogram in dB (middle), and phase spectrogram (right) of female speech. While the magnitude spectrum presents a detailed harmonic structure of speech in time and frequency, the instantaneous phase spectrum shows no useful pattern or useful details.

contributions over time. The magnitude part reveals a clear harmonic structure, and has been shown to follow a heavy-tailed distribution, e.g., super-Gaussian (see, e.g., Vary and Martin 2006). Information about formants and harmonics are mixed, and the overall presentation is redundant due to the large dimensionality of the dictionary (basis functions) used in STFT to present the sparse speech signal.

The phase spectrum shown in the right panel, shows a random pattern with no specific harmonic structure, unlike the magnitude part, and hence is not accessible for the analysis of a speech signal. The phase spectrum is in particular often reported to follow a uniform distribution $p_\Phi(\phi) = \frac{1}{2\pi}$ with $\phi \in [-\pi, \pi[$ known as the maximum uncertainty. Therefore, spectral phase, in contrast to its magnitude counterpart, reveals no visible or useful structure. This is due to the phase wrapping phenomenon. More details about the phase wrapping problem and possible ways to resolve it are discussed in Section 2.3.

In the authors' opinion, the unavailability of a useful STFT phase structure has been the main reason why relatively less research has been conducted towards phase-based signal processing compared to its magnitude counterpart, to which a large amount of literature has been dedicated. To address this point, in Section 2.4 we will present several contributions where useful phase representations provide accessible spectral phase analysis.

2.3 Phase Unwrapping

2.3.1 Problem Definition

The phase spectrum extracted from the discrete-time Fourier transform may vary by multiples of 2π due to its periodic nature. This leads to discontinuities, and hence the phase spectrum becomes non-unique. This problem is called *phase wrapping* and is a key problem in many signal processing applications, including interferometry (Goldstein *et al.* 1988), synthetic aperture radar, field mapping in magnetic resonance imaging (MRI; Ying 2006), wavefront distortion measurement in adaptive optics, accurate profiling of mechanical parts by X-ray, and finally in speech and image processing (Oppenheim and Lim 1981).

In general, for a given frequency of interest ω_k, the wrapping problem refers to the fact that the values for the desired unwrapped phase, i.e., the continuous phase function denoted by $\arg[X(e^{j\omega})]$ differs from the principal value of the phase up to unknown jumps of 2π,

$$\arg[X(e^{j\omega_k})] = \text{ARG}[X(e^{j\omega_k})] + 2\pi m_k, \tag{2.4}$$

where the correct value of $m_k \in \mathbb{Z}$ is not known and should be chosen to satisfy the continuity condition. Further, we define $\text{ARG}[X(e^{j\omega})]$ as the principal value given by:

$$\text{ARG}[X(e^{j\omega_k})] = \tan^{-1}\left[\frac{X_{\text{Im}}(e^{j\omega_k})}{X_{\text{Re}}(e^{j\omega_k})}\right], \tag{2.5}$$

where $X_{\text{Re}}(e^{j\omega})$ and $X_{\text{Im}}(e^{j\omega})$ are the real and the imaginary parts of the spectrum. The angle for the signal Fourier transform $\angle X(e^{j\omega})$ at each frequency is therefore only unique up to an additive unknown multiple of 2π. For speech signals as real signals, the magnitude is even, while the phase is odd following the conjugate symmetric property of the spectrum. For the case of a real input signal, the complex cepstrum is also real, implying that the unwrapped phase should be odd symmetric.[1] For more details on the complex cepstrum and its properties, we refer to Oppenheim and Schafer, (1989).

In order to visualize the phase unwrapping problem in the STFT, we consider the following example. The signal $x(n)$ is a speech waveform for the vowel "e" sampled at $f_s = 16$ kHz. Figure 2.2 visualizes the result of this experiment: (a) time domain waveform, (b) STFT magnitude representation, where a frame shift of one sample together with a frame length of 512 samples (32 ms) is used. The black dashed lines mark the trajectories of the fundamental frequency of the first three harmonics, referred to as $h = 1, 2, 3$. A causal analysis window (c) introduces a linear phase that covers any useful harmonic pattern in the spectral phase. In contrast, using an acausal, symmetric window with zero phase, some phase structure can be revealed (d).

Figure 2.2 further demonstrates the harmonic representation of the phase spectrum. In (e), the waveform is presented as a sum of harmonics where the underlying sinusoids for the first three harmonics are shown. The superposition of the three harmonics is shown in (f), while their instantaneous phase is shown in (g). The harmonics are estimated by applying the Fourier transform along the trajectories of the fundamental frequency (the black dashed line shown in (b) and (d)). The figure reveals that the instantaneous phase in (g) provides less information and the strong wrapping impact on phase is visible. Knowledge of the fundamental frequency allows an efficient unwrapping of the harmonic phases, disclosing the structure of the unwrapped phase (h). We note here that (h) is obtained by phase unwrapping along time, while (d) is achieved along frequency.

Let z_m be the mth zero in the z-transform of a finite-length discrete real signal $x(n)$ with $z = e^{j\omega}$. Figure 2.3 illustrates the main difficulty in the existing DFT-based phase unwrapping solutions due to the zeros close to the unit circle. The extent of the phase wrapping depends on the location of the zeros in the z-plane relative to the unit circle, as well as the value of the linear phase component (Murthy and Yegnanarayana 1991a). The closer the zeros of the z-transform are to the unit circle, the sharper the phase change at frequencies in their vicinity. As a consequence, substantial jumps in phase are possible when the resolution of the discrete Fourier transform is not sufficiently large. In particular, the phase jump experienced between ω_{k-1} and ω_k could be large if the mth zero z_m is close enough to the unit circle. These sharp changes are the main reason for

---

1 An odd symmetric signal is symmetric about the origin.

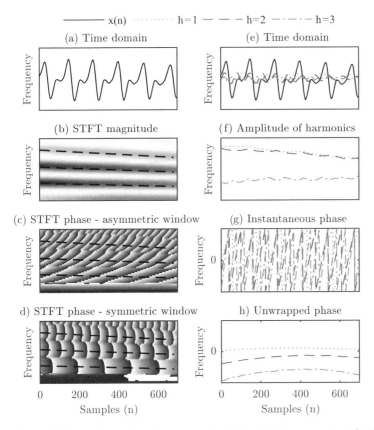

—— x(n) ·········· h=1 — — — h=2 —·—·—· h=3

(a) Time domain   (e) Time domain

(b) STFT magnitude   (f) Amplitude of harmonics

(c) STFT phase - asymmetric window   (g) Instantaneous phase

d) STFT phase - symmetric window   h) Unwrapped phase

Figure 2.2 Example showing phase wrapping in the STFT phase spectrum for the vowel "e": (a) $x(n)$, waveform; (b) $\phi(k,l)$, STFT magnitude $|X(k,l)|$; (c) STFT phase using a causal window; (d) $\phi'(k,l)$, STFT phase using an acausal symmetric window with zero phase; (e) waveform representation as a sum of harmonics; (f) $A(h,l)$, amplitude of the hth harmonic; (g) $\psi(h,l)$, harmonic instantaneous phases; (h) $\Psi(h,l)$, unwrapped phase.

Figure 2.3 An example for two zeros close to the unit circle $|z| = 1$ located between ω_{k-1} and ω_k. Such zeros are the main source of difficulty for DFT-based phase unwrapping methods (Drugman and Stylianou 2015).

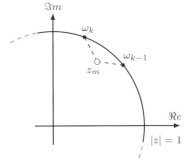

the difficulty in phase unwrapping. As we will see throughout the next section, in order to avoid this problem, a relatively large number of DFT samples (high DFT resolution) is required.

2.3.2 Remarks on Phase Unwrapping

The goal in a *phase unwrapping* algorithm is to find m_k in (2.4), to ensure a continuous phase function. The phase unwrapping problem is an ill-conditioned one for the following reasons (Quatieri 1979):

- The existing phase unwrapping solutions are prone to errors when the sampling frequency is not large enough. Similarly, errors occur when the Fourier transform involves regions where the signal energy is low, which are susceptible to degradation due to quantization noise.
- Components with slowly varying characteristics in the unwrapped phase (e.g., the envelope of the signal) are prone to small changes of a sequence in the time domain. Such high sensitivity contributes to an ill-conditioned problem.

The principal value of the phase, is obtained by constraining the phase values to lie in the range $[-\pi, \pi[$ for any ω, hence forcing

$$- \pi < \text{ARG}[X(e^{j\omega})] \leq \pi. \tag{2.6}$$

The following constraints are required for an unwrapped phase function to exist (Karam 2006):

1) It is required to be a continuous function of frequency ω. This condition constrains the z-transform $X(z)$ to have no zeros on the unit circle. This leads to a discontinuity of π in the phase evaluated at frequency ω, thus violating the continuity condition.
2) Its z-transform $X(z)$ should have no linear phase term of the form z^r, as they lead to discontinuities in the unwrapped phase at $\omega = 2\pi$. Removing linear phase terms is possible either by subtracting out the linear term or by properly shifting the input sequence (Quatieri 1979).
3) We define the DTFT of the complex cepstrum as:

$$\hat{X}_{cc}(e^{j\omega}) = \log|X(e^{j\omega})| + j\arg[X(e^{j\omega})]. \tag{2.7}$$

   The imaginary part in the complex cepstrum is ambiguous since it is unique only up to an additive multiple of 2π at frequency ω. In order to have a valid DTFT for the complex cepstrum the unwrapped phase function $\arg[X(e^{j\omega})]$ should be periodic with a period of 2π, as well as a continuous and odd function. It is important to note that, as shown by Tribolet (1977), the unwrapped phase is odd only when the mean of the phase derivative $\frac{1}{2\pi}\int_0^{2\pi}\arg'[X(e^{j\omega})]/d\omega$ is equal to zero. If the mean value is not equal zero, linear phase removal must be carried out after unwrapping.
4) For a real-valued signal the complex cepstrum is also real; therefore, the discrete short-time Fourier transform will be conjugate symmetric.

2.3.3 Phase Unwrapping Solutions

Several researchers have made attempts to solve the phase wrapping problem. The works in Ghiglia and Pritt (1998) and Ying (2006) reviewed two-dimensional phase unwrapping

Table 2.1 List of phase unwrapping solutions.

Phase Unwrapping Method	Reference
Detecting discontinuities (DD)	Oppenheim and Schafer (1989)
Numerical integration (NI)	Tribolet (1977); Bonzanigo (1978)
Isolating sharp zeros	Nashi (1989)
Iterative phase unwrapping	Quatieri and Oppenheim (1981)
Polynomial factorization (PF)	Steiglitz and Dickinson (1982); Sitton *et al.* (2003)
Time series	McGowan and Kuc (1982); Long (1988); Keel and Bhattacharyya (2002)
Composite method	Karam and Oppenheim (2007)
Schur–Cohn and Nyquist frequency	Drugman and Stylianou (2015)

algorithms. Together with the theoretical aspects, implementations of each algorithm were presented. As the problem of image reconstruction using phase unwrapping falls beyond the scope of this book, in the rest of this chapter we will focus on phase unwrapping methods that were developed for one-dimensional data, in particular for speech signals. Table 2.1 lists the well-known solutions for phase unwrapping available in the literature that will be discussed in the following.

2.3.3.1 Detecting Discontinuities

As a direct solution for phase unwrapping, Oppenheim and Schafer (1989) proposed a phase unwrapping method called *detecting discontinuities (DD)*. The idea was to look for discontinuities and to detect and remove them by checking the differences in phase between the adjacent samples of the principal values, reaching beyond π. Any value detected to have a change larger than π will get wrapped to lie in the acceptable range for phase, i.e., $[-\pi, \pi[$.

The goal is to find the integer multiple of 2π for each frequency ω_k, expressed as $2\pi m_k$, that when added to the principal phase function $\mathrm{ARG}[X(e^{j\omega})]$ yields a continuous phase $\phi(e^{j\omega_k})$, as given in (2.4). Assuming that $m_0 = 0$, m_k can be found with the following steps (Quatieri 2001):

1) A jump of $+2\pi$ is detected if $\mathrm{ARG}[X(e^{j\omega_k})] - \mathrm{ARG}[X(e^{j\omega_{k-1}})] > 2\pi - \varepsilon$. Then we subtract 2π, i.e., $m_k = m_{k-1} - 1$.
2) A jump of -2π is detected if $\mathrm{ARG}[X(e^{j\omega_k})] - \mathrm{ARG}[X(e^{j\omega_{k-1}})] < -(2\pi - \varepsilon)$. Then we add 2π, i.e., $m_k = m_{k-1} + 1$.
3) In the case of no discontinuity detection we have $m_k = m_{k-1}$.

The method is fast and simple. It yields correct unwrapped phase estimates if the frequency spacing is small enough; in particular, the distance between two adjacent samples in the unwrapped phase should be less than the threshold ε. However, due to possible zeros close to the unit circle, rapid changes in phase are inevitable. In particular, for such zeros a phase change of $-\pi$ or π between frequencies ω_{k-1} and ω_k occurs when it lies inside or outside the unit circle, respectively. Therefore, to avoid errors in the unwrapping process, a high resolution of the DFT is required. On the other hand, in order to have a good balance between computational load (DFT length) and unwrapping accuracy the DD approach has been extended by the matching method. Starting from a low DFT resolution, the resolution is increased until the unwrapped phase estimates of two successive resolutions are equal. The method is called the *matching* (M) method as an extension of the DD method.

The DD method is implemented in the *unwrap* function in MATLAB® and is limited as it assumes a maximum change of π from one frequency sample to the next, which is not valid in non-stationary changes in a real speech signal. Further details on how the method works on clean or noisy signals will be presented in Experiment 2.1 in Section 2.5.1.

2.3.3.2 Numerical Integration (NI)

This method assumes that the principal value is given by integrating the uniquely defined phase derivative $\phi'(e^{j\omega}) = \frac{d\phi(e^{j\omega})}{d\omega}$, and we have:

$$\phi(e^{j\omega_k}) = \int_0^{\omega_k} \phi'(e^{j\omega})d\omega. \tag{2.8}$$

The unwrapped phase given in (2.8) cannot be precisely calculated in discrete time. Therefore, Tribolet (1977) proposed to approximate the unwrapped phase by *adaptive numerical integration* (ANI). As numerical errors accumulate in the integral, significant errors might occur. As a result, to circumvent this issue and to reduce the resulting errors, information about the principal phase value was incorporated. Assuming that the value of the unwrapped phase at frequency ω_{k-1} is known, using the trapezoidal approximation for numerical integration, an estimate for the unwrapped phase at frequency ω_k is given by:

$$\hat{\phi}(e^{j\omega_k}) = \phi(e^{j\omega_{k-1}}) + \frac{\omega_k - \omega_{k-1}}{2}\left(\left.\frac{d\phi(e^{j\omega})}{d\omega}\right|_{\omega=\omega_k} + \left.\frac{d\phi(e^{j\omega})}{d\omega}\right|_{\omega=\omega_{k-1}}\right). \tag{2.9}$$

Then the optimization to find an estimate for m_k (denoted by $\hat{m}_k$) is formulated as minimizing the difference between the predicted value and the candidate value over all possible m_ks, and we have:

$$\hat{m}_k = \arg\min_{m_k} \varepsilon_k, \tag{2.10}$$

$$\varepsilon_k = |\text{ARG}[X(e^{j\omega_k})] + 2\pi m_k - \hat{\phi}(e^{j\omega_k})|, \tag{2.11}$$

where $\hat{\phi}(e^{j\omega_k})$ is given from (2.9). The method performs well when the zeros are sharp and separate enough, which requires a small enough $\omega_k - \omega_{k-1}$ as the increment step.

Later, Bonzanigo (1978) proposed a modified version of Tribolet's solution to obtain improved phase unwrapping performance. Although Tribolet's phase unwrapping solution led to good results, the execution time for the algorithm is long, in particular when the phase derivative becomes irregular with large peaks. To improve the efficiency of Tribolet's algorithm, Bonzanigo suggested modifying the algorithm by replacing the DFT method with a modified Goertzel algorithm (Goertzel 1958). The Goertzel algorithm is extended by a phase correction term suggested to interpolate between DFT points.

2.3.3.3 Isolating Sharp Zeros

Nashi (1989) proposed to take into account the radial position of the sharp zero with respect to the unit circle to compute the unwrapped phase. The method works in segments of the z-plane, defined as the frequency step between two consecutive DFT samples of $\frac{2\pi}{N_{\text{DFT}}}$, with N_{DFT} denoted as the DFT length. Particular attention is paid to sharp zeros, defined as zeros that lie between the radial lines of 0.99 and 1.01, and hence are close to the unit circle. Such information about the zeros is useful to determine the *phase changes* (PC). This was used to calculate an unwrapped phase, explained in the following.

The algorithm applies two measures to detect the sharp zeros:

- *Quasi-phase derivative with respect to the radial distance* (QPDR) at the frequency sampling points,

$$\text{QPDR} = Q_1 \cdot (PC)/r_s, \tag{2.12}$$

where r_s is the radial step size, PC is the phase change between the trailing and leading segment due to the presence of m zeros, and Q_1 is a constant.
- *Phase increment* (PHINC) at the frequency sampling points due to an incremental radial expansion step, defined as

$$\text{PHINC} = |\text{PHADER}|$$
$$= |Q_2 \cdot (\text{phase change at a point})/r_s|, \tag{2.13}$$

where $\text{PHADER} = -\frac{d\phi_1}{dr}$ is the phase derivative at any frequency sampling point k with respect to the radial distance r, ϕ_1 is the respective phase angle due to the zero at the trailing end of the segment, and Q_2 is a constant.

The radial positions detected in this way are then used to quantify the PC across the detected segment. By detecting sharp zero locations, the phase unwrapping allows computing the predicted phase change across a segment. Comparing this predicted phase change outcome with the wrapped phase change obtained directly from the DFT, m_k can be estimated by searching for the integer value which brings these two phase change values closest. The method performs well for the case of a small number of sharp zeros.

2.3.3.4 Iterative Phase Unwrapping

Quatieri and Oppenheim (1981) proposed an iterative solution to find the unwrapped phase of the DTFT of an input sequence $x(n)$. Their method relied on two important properties:

- Given a sequence $x(n)$ with its z-transform denoted by $X(z)$, the minimum phase property requires $x(n)$ to be causal, i.e., $x(n) = 0$ for $n < 0$ and $x(0) = A_0$, where A is the scale factor in the z-transform $X(z)$ given in (2.15). Under these conditions, the resulting unwrapped phase function will have no linear phase component.
- The log-magnitude and unwrapped phase of $X(e^{j\omega})$ denoted by $\arg[X(e^{j\omega})]$ is related through the Hilbert transform, and we have:

$$\arg[X(e^{j\omega})] = \phi(e^{j\omega}) = \mathcal{H}(\ln|X(e^{j\omega})|), \tag{2.14}$$

where $\mathcal{H}(\cdot)$ denotes the Hilbert transform and $\ln|X(e^{j\omega})| = \log_e|X(e^{j\omega})|$.

The iterative phase unwrapping approach works in the following steps:

1) Remove the linear phase and calculate the minimum-phase sequence from the principal value of the phase of the input sequence.
2) Apply the Hilbert transform using the relationship in (2.14) on the minimum-phase sequences (Oppenheim and Schafer 1989).
3) Finally, the unwrapped phase spectrum is calculated by adding back the linear phase part (obtained in the first step) to the estimated minimum phase (from the second step).

This method, although effective on short data lengths, requires *a priori* knowledge about the linear phase as well as a large DFT size within iterations to avoid aliasing. Both requirements restrict the performance and successful results were reported for short sequences only.

2.3.3.5 Polynomial Factorization (PF)

Steiglitz and Dickinson (1982) proposed a phase unwrapping algorithm relying on the following polynomial factorization principle. Let $X(z)$ be the z-transform for the finite-length sequence $x(n)$ which is given by:

$$X(z) = A_0 z^r \prod_{k=1}^{N_{IUC}} (1 - a_k z^{-1}) \prod_{k=1}^{N_{OUC}} (1 - b_k z^{-1}), \tag{2.15}$$

with N_{IUC} and N_{OUC} denoting the number of zeros inside and outside the unit circle, respectively, and A_0 is a scalar. The term z^r in (2.15) contributes as a linear phase term. For a causal signal we have $r = -N_{OUC}$ and the phase of the signal is given by:

$$\arg[X(e^{j\omega})] = \sum_{k=1}^{M_I} \arg[(1 - a_k e^{-j\omega})] + \sum_{k=1}^{M_0} \arg[(1 - b_k e^{-j\omega})]. \tag{2.16}$$

Then the polynomial factorization phase unwrapping (Steiglitz and Dickinson 1982) suggests first finding the roots of $X(z)$ and then calculating the phase contribution of each of the roots as detailed in (2.16). The unwrapped phase is given as the superposition of the phase contribution of all the roots in the polynomial.

As sampled physical signals tend to have their zeros close to the unit circle, therefore, the effectiveness of phase unwrapping becomes restricted. It was also reported that for a sufficiently fine grid size, the outcome of Tribolet's adaptive numerical integration technique (Tribolet 1977) coincides with that obtained by the polynomial factorization method, when the numerical errors are neglected.

Later, Sitton *et al.* (2003) proposed an efficient version for the polynomial factorization approach by using the fast Fourier transform to establish a search grid around the unit circle. The z-transform of the signal was then evaluated at evenly spaced samples lying on concentric circles with radii close to 1.

2.3.3.6 Time Series Approach

McGowan and Kuc (1982) proposed a phase unwrapping approach relying on the discrete-time sequence. The algorithm calculates the integer-valued function $l(\omega_k)$, which results in a continuous phase by adding or subtracting an integer multiple of π.

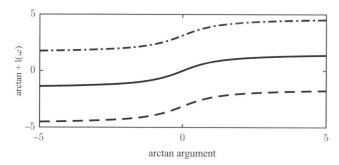

Figure 2.4 Different branches for the arctan function are used in McGowan and Kuc (1982) to determine $l(\omega)$ in (2.13) for adding or subtracting the π multiples required in the time series phase unwrapping method.

Then the phase at some frequency $0 \le \omega_1 \le \pi$ relative to $\omega = 0$ is given by (McGowan and Kuc 1982):

$$\phi(e^{j\omega_1}) - \phi(e^{j\omega_0}) = - \left\{ \arctan \left(\frac{X_\mathrm{I}(e^{j\omega_1})}{X_\mathrm{R}(e^{j\omega_1})} \right) + \pi l(\omega_1) \right\}. \tag{2.17}$$

Determining the integer $l(\omega_1)$ is equivalent to solving the phase unwrapping process. The addition or subtraction by π depends on the sign change in the ratio of the arctan argument, i.e., $\frac{X_\mathrm{I}(e^{j\omega})}{X_\mathrm{R}(e^{j\omega})}$. As the product $X_\mathrm{I}(e^{j\omega})X_\mathrm{R}(e^{j\omega})$ has the same sign as the ratio, they further considered the product due to its advantage of assuring non-singularity. Further, the value of $l(\omega)$ is ascertained by tracking the number of times that the sign of the arctan outcome is changed when ω goes through zeros. The phase unwrapping is converted to determine the sign changes of the product term $X_\mathrm{I}(e^{j\omega_1})X_\mathrm{R}(e^{j\omega})$ by finding the roots of $X_\mathrm{R}(e^{j\omega})$. The branches of the arctan function are shown in Figure 2.4.

The method calculates the unwrapped phase at any frequency directly from the time series without any intermediate value. The method, however, is limited to quite short time sequences. Long (1988) extended the method for longer time sequences; however, the method was reported to be less accurate due to the increased computational errors. Finally, Keel and Bhattacharyya (2002) proposed a phase unwrapping method relying on the root distribution with respect to the unit circle, which was determined using Tchebyshev polynomials.

2.3.3.7 Composite Method

Karam and Oppenheim (2007) proposed two phase unwrapping solutions where previous methods were combined to compensate for their shortcomings. The motivation was to benefit from the combination of the strengths of the DD, ANI, and PF approaches, while avoiding their weaknesses. Their combined method relied on decoupling the signal into two terms, one part with zeros problematic for DD and ANI, and the other part containing the remaining zeros. The steps in the composite method were as follows:

1) The method starts with polynomial factorization to find the zeros near the unit circle, and calculates the unwrapped phase corresponding to the contribution of these zeros.

2) By removing the zeros detected in the first step, another polynomial of a lower degree is produced.

3) The unwrapped phase is calculated for the lower-degree polynomial found in the previous step, using either DD or ANI.

4) Finally, the overall unwrapped phase is obtained by adding the unwrapped phase parts estimated in the first and the third steps for the individual polynomials.

The combined method was shown to outperform each individual method evaluated on synthetic signals (Karam 2006; Karam and Oppenheim 2007).

2.3.3.8 Schur–Cohn and Nyquist Frequency

Drugman and Stylianou (2015) presented an overview of phase unwrapping solutions in the DFT domain. They also proposed a phase unwrapping method relying on the link between the unwrapped phase value at the Nyquist frequency $\omega = \pi$ given by:

$$\phi(e^{j\phi}) = \pi(-N + 1 + N_{\text{IUC}}) = -\pi N_{\text{OUC}}, \tag{2.18}$$

where N_{IUC} and N_{OUC} denote the number of zeros inside and outside the unit circle, respectively. A fast calculation for the root distribution of polynomial roots is possible using their modified version of the Schur–Cohn algorithm. Having initialized for N_{OUC} and the number of DFT points, an estimated unwrapped phase is calculated using the DD technique. Then, from (2.18), N_{OUC} is updated to $-\frac{\phi(e^{j\phi}) - \phi(e^{j0})}{\pi}$. The procedure is iterated by doubling the value of the DFT points until no change in the N_{OUC} estimate is observed.

The method is computationally less expensive compared to the other state-of-the-art phase unwrapping techniques described earlier in this section. The results of experiments conducted on datasets consisting of synthetic random and real speech signals demonstrate that their proposed method outperforms in terms of estimation accuracy of the unwrapped phase at a reduced computational cost.

In Experiment 2.1 we will present examples for phase unwrapping of a one-dimensional signal followed by a comparative study to evaluate the performance of the phase unwrapping methods described in this chapter.

2.4 Useful Phase-Based Representations

In the following, we will focus on several useful phase-based features which, in contrast to the instantaneous phase extracted from Fourier analysis, are helpful to get further insights into the spectral phase of a signal. Table 2.2 presents a list of the phase representations discussed in this section: instantaneous STFT phase, *group delay* (GD) and its various extensions (Yegnanarayana and Murthy 1992; Stark and Paliwal 2009; Hegde *et al.* 2007; Bozkurt *et al.* 2007), *instantaneous frequency* (IF) (Boashash 1992a; Stark and Paliwal 2008), harmonic phase (McAulay and Quatieri 1986), baseband phase difference (Krawczyk and Gerkmann 2014), phase variance (Fisher 1995), *phase distortion standard deviation* (PDD) (Degottex *et al.* 2011), and *relative phase shift* (RPS; Saratxaga *et al.* 2009). For each phase representation, the list shown in Table 2.2 also

Table 2.2 List of useful phase representations explained in this chapter.

Representation	Remarks
Group delay ($\tau(\omega)$)	Noise robust, good formant representation (Yegnanarayana and Murthy 1992)
Instantaneous frequency (IF(k, l))	Captures signal's non-stationarity (Boashash 1992a)
Harmonic phase ($\phi(h, l)$)	Speech analysis/synthesis (McAulay and Quatieri 1986; Degottex *et al.* 2011)
Phase distortion (PD(h, l))	Glottal shape estimation (Degottex *et al.* 2011, 2012)
Phase variance ($\sigma_c(h, l)$)	Voice quality assessment (Koutsogiannaki *et al.* 2014; Degottex and Obin 2014)
Relative phase shift (RPS(h, l))	Speech synthesis (de Leon *et al.* 2011); speaker recognition (Sanchez *et al.* 2015)
Phasegram: unwrapped harmonic phase (Ψ)	Speech synthesis (Degottex and Erro 2014); enhancement (Mowlaee and Kulmer 2015b)
Baseband phase difference (BPD)	Improved STFT phase reconstruction (Krawczyk and Gerkmann 2014)

summarizes some useful properties and corresponding references where the representation has been successfully used. In the following, we present the concept used in each phase representation. Later, we show successful usage of each representation in the context of speech signal processing.

2.4.1 Group Delay Representations

Group delay is perhaps one of the most useful and well-studied phase representations in speech signal processing. Group delay is defined as the negative first derivative of the phase spectrum with respect to frequency, given by:

$$\tau(\omega) = -\frac{d\phi(e^{j\omega})}{d\omega}. \tag{2.19}$$

For discrete frequencies, the differentiation can be approximated by

$$\tau(k) = -\frac{\phi(e^{j\omega_{k+1}}) - \phi(e^{j\omega_k})}{\omega_{k+1} - \omega_k}, \tag{2.20}$$

defined for consecutive frequencies ω_k and ω_{k+1}. This de-emphasizes the wrapping of the Fourier spectral phase described earlier in this chapter (see Section 2.3). As a result, certain useful harmonic structure in the speech signal becomes visible in the group delay representation (see, for example, the group delay pattern shown in Figure 2.16).

Group delay is a popular representation used by researchers in different applications, such as waveform estimation (Yegnanarayana *et al.* 1985), signal reconstruction (Yegnanarayana *et al.* 1984), and spectrum estimation (Yegnanarayana and Murthy 1992). It has also been successfully used as a viable analysis tool in a variety of applications in speech signal processing, such as formant extraction (Murthy and Yegnanarayana 1991a), speech segmentation (Prasad *et al.* 2004), voice activity detection (Parthasarathi *et al.* 2011), and improved signal reconstruction (Mowlaee *et al.* 2012; Mowlaee and Saeidi 2014).

To avoid the approximation in (2.20), starting from the definition of group delay $\tau(\omega)$ given in (2.19) and the fact that $\ln(X(e^{j\omega})) = \ln|X(e^{j\omega})| + j\phi(e^{j\omega})$ we get:

$$
\begin{aligned}
\tau(\omega) &= -\text{Im}\left(\frac{d\ln(X(e^{j\omega}))}{d\omega}\right) \\
&= -\text{Im}\left(\frac{\frac{d}{d\omega}(\sum_{n=-\infty}^{\infty} x(n)e^{-j\omega n})}{\sum_{n=-\infty}^{\infty} x(n)e^{-j\omega n}}\right) \\
&= \text{Im}\left(j\frac{X^*(e^{j\omega}) \cdot Y(e^{j\omega})}{X^*(e^{j\omega}) \cdot X(e^{j\omega})}\right),
\end{aligned}
\tag{2.21}
$$

defining $Y(e^{j\omega}) = \text{DTFT}\{nx(n)\}$ and $\text{Im}(\cdot)$ as the operator returning the imaginary part. Using the real and imaginary parts of X and Y, i.e., $X = X_R + jX_I$ and $Y = Y_R + jY_I$ we finally get:

$$
\tau(\omega) = \frac{X_R(e^{j\omega}) \cdot Y_R(e^{j\omega}) + X_I(e^{j\omega}) \cdot Y_I(e^{j\omega})}{|X(e^{j\omega})|^2},
\tag{2.22}
$$

where $Y(e^{j\omega})$ is the Fourier transform of the sequence $nx(n)$, and the subscripts R and I denote the real and imaginary parts of the complex coefficients. From the group delay representation in (2.22) it is obvious that when the magnitude spectrum gets close to zero, the group delay function gets spiky, which limits its usefulness. In particular, excitation components introduce these zeros, and hence restrict the applicability of the group delay.

Group delay representation has been reported to be robust against additive noise (see, e.g., the analysis given in Parthasarathi *et al.* 2011). Its robustness makes it a favorable candidate as an auxiliary feature to improve the performance in certain speech processing applications, for example *automatic speech recognition* (Hegde *et al.* 2007), speaker identification (Hegde *et al.* 2004), and sound classification (Kandia and Stylianou 2008). In particular, working with the group delay function has several advantages (Loweimi *et al.* 2015): (i) it has a high frequency resolution, which enables it to resolve the closely located peaks in the speech spectrum; (ii) under certain conditions it presents a clear behavior which resembles the magnitude spectrum; and (iii) it is additive for two signals which are convolved in the time domain.

The zeros of a signal's z-transform may mask the formants and the desired harmonic structure in the spectrum due to introducing some spurious peaks in the group delay. As a result, the group delay spectrum could get unstable. To avoid these spurious peaks in the group delay representation, several different solutions have been proposed in the literature. Table 2.3 lists the methods proposed to resolve this issue whereby a modified version for the group delay function has been proposed; these are reviewed below.

Hegde *et al.* (2007) proposed to cepstrally smooth the spectral amplitude $|X(e^{j\omega})|$ in the denominator of the group delay $\tau(\omega)$. The resulting cepstrally smoothed spectrum is denoted by $S(e^{j\omega})$, and the corresponding modified smoothed group delay function denoted by $\tau_m(\omega)$ is given by (Hegde *et al.* 2007):

$$\tau_s(\omega) = \frac{X_R(e^{j\omega})Y_R(e^{j\omega}) + Y_I(e^{j\omega})X_I(e^{j\omega})}{|S(e^{j\omega})|^{2\gamma}}, \tag{2.23}$$

$$\tau_m(\omega) = \left(\frac{\tau_s(\omega)}{|\tau_s(\omega)|}\right)|\tau_s(\omega)|^{\alpha_0}, \tag{2.24}$$

where $\tau_s(\omega)$ refers to a smoothed group delay version, applying the tuning parameters γ and α_0. Hegde *et al.* showed improved performance in automatic speech and speaker recognition applications when using the modified group delay features as their selected features.

As another modification of the group delay representation, Bozkurt *et al.* (2007) proposed the *chirp group delay* (CGD), defined as the negative derivative of the phase spectrum (the group delay function) computed from the chirp z-transform. Given the chirp z-transform (Rabiner *et al.* 1969), the chirp group delay is given by taking the frequency derivative. The z-analysis on a circle rather than the unit circle de-emphasizes the spikes originating from the zeros of the z-transform which mask formant peaks in the group delay function. This important benefit results in improved ASR performance (Bozkurt 2005).

To reduce the negative effects on the group delay spectrum due to zeros on the unit circle, Alsteris and Paliwal (2007) proposed an alternative group delay spectrum where cepstral smoothing is applied. Alternatively, Bozkurt and Couvreur (2005) applied the Mel-scale filterbank on the group delay representation followed by cepstral transformation. Zhu and Paliwal (2004) proposed multiplying the group delay and power spectrum to obtain a product spectrum given by

$$X_{PS}(e^{j\omega}) = |X(e^{j\omega})|^2 \tau(e^{j\omega}) = X_R(e^{j\omega})Y_R(e^{j\omega}) + Y_I(e^{j\omega})X_I(e^{j\omega}). \tag{2.25}$$

The choice of window type and length is a crucial factor in the accuracy of the group delay calculation (Bozkurt and Couvreur 2005). A proper selection of the analysis window contributes in resolving the formant structure in the group delay representation. Zeros close to the unit circle in the z-plane result in spikes in the group delay function, and consequently mask the vocal tract information presented by the group delay (Murthy and Yegnanarayana 1991b). In Rajan *et al.* (2013), the group delay function was calculated using the phase of all-pole modeling rather than the Fourier spectrum. The new feature was called *linear prediction group delay* (LPGD). Loweimi *et al.* (2013) proposed to couple the group delay function with an autoregressive model which helps to improve the frequency resolution of the estimated power spectrum with a low frequency leakage. The group delay feature representation leads to improved automatic speech recognition compared to the standard MFCCs.

Finally, Hideki *et al.* (1998) proposed a *time domain smoothed group delay* (TSGD), which benefits from the advantageous glottal shape property of the group delay spectrum. Preserving this waveform information results in producing speech signals almost indistinguishable from the originals. Similarly, in Kawahara and Masuda (1996) a channel vocoder called STRAIGHT was proposed to employ minimum phase for speech synthesis. Kawahara and Masuda showed that the group delay information plays a vital role in delivering a high quality synthesized speech signal.

Table 2.3 Group delay functions and variants.

Group Delay Representation	Reference
Group delay deviation	Stark and Paliwal (2009)
Cepstrally smoothed group delay spectrum	Bozkurt and Couvreur (2005)
Chirp group delay (CGD)	Bozkurt *et al.* (2007)
Modified group delay (MGD)	Hegde *et al.* (2007)
Time domain smoothed group delay (TSGD)	Hideki *et al.* (1998)
Linear prediction group delay (LPGD)	Rajan *et al.* (2013)

Stark and Paliwal (2009) introduced a group-delay-based feature representation and showed its effectiveness for speech analysis purposes. The new feature relied on the deviation of the group delay for each sinusoid. Let $w(n)$ be the *finite impulse response* (FIR) window with non-zero values within $n \in [0, N_w - 1]$. The window function introduces a linear phase response, and hence its resulting group delay is constant and equal to $\tau_w = \frac{N_w - 1}{2}$. The *group delay deviation* (GDD) is defined as the deviation in group delay of $\tau_x(k, l)$ with respect to τ_w, given as (Stark and Paliwal 2009)

$$\Delta \tau_x(k, l) = \tau_w - \tau_x(k, l). \tag{2.26}$$

Stark and Paliwal found that the group delay deviation obtained from the phase spectrum of a clean speech signal exhibits minima at spectral harmonics; therefore, a smooth group delay trajectory is expected at the locations of the signal harmonics (see panels (c) and (d) in Figure 2.2). This property of group delay was further used as the additional constraint to resolve the ambiguity in the phase estimation problem for signal reconstruction in single-channel source separation (Mowlaee *et al.* 2012) and in speech enhancement (Mowlaee and Saeidi 2014). For further details, see the geometry-based phase estimators explained in Chapter 3.

2.4.2 Instantaneous Frequency

To represent the temporal behavior of the spectral phase, an alternative phase representation is the instantaneous frequency (IF). Here, we first briefly provide the IF definition in the continuous domain and proceed with its discrete-time variant. Consider a single sinusoid defined by its amplitude a, frequency f, and time-varying phase term denoted by $\theta(t)$:

$$x(t) = a \cos(2\pi f t + \theta(t)). \tag{2.27}$$

The IF is defined as the first time derivative of the instantaneous phase given by (Carson and Fry 1937)

$$\omega_i(t) = \frac{d}{dt}(2\pi f t + \theta(t)) = 2\pi f + \frac{d\theta(t)}{dt}, \tag{2.28}$$

interpreted as the rate of change of the phase angle at time instant t. In the discrete domain the IF for two consecutive time samples with frame shift S is defined by:

$$\omega_{\text{IF}}(k, l) = \frac{\text{ARG}[X(k, l)X^*(k, l - 1)]}{S}, \tag{2.29}$$

where $\omega_{\mathrm{IF}}(k, l)$ denotes the instantaneous frequency calculated at frequency bin k and frame index l. The IF representation captures the non-stationarity characteristics of a signal compared to the center frequency represented by the STFT. A fixed frequency resolution is chosen, hence it is limited to the primary choice of the prototype window (see, e.g., the detailed discussion presented in Chapter 4 of Vary and Martin 2006). For speech, the IF representation provides more information than the spectrogram, in particular about the transients, as reported in Cohen (1989).

As an alternative for IF, the *instantaneous frequency deviation* (IFD) has been defined in the literature (Friedman 1985; Stark and Paliwal 2008):

$$\mathrm{IFD}(k, l) = \mathrm{ARG}[X(k, l)X^*(k, l-1)e^{-j\omega_k}], \tag{2.30}$$

where * denotes the complex conjugate. The IF tracks the harmonic frequency accurately as the corresponding spectral magnitude increases; hence, the IFD is inversely related to the spectral magnitude, as shown in (2.30). McCowan *et al.* (2011) reported successful automatic speech recognition using IFD as extracted features.

The IF representation has been successfully employed in speech processing applications, for example pitch estimation (Stark and Paliwal 2008), where the IFD spectrum was reported useful in representing both pitch and formant structures in speech, and in estimation theory where several robust estimators of IF for sinusoidal components observed in noise have been reported. As an example, Lagrange and Marchand (2007) used the temporal derivative of phase to approximate IF. A maximum likelihood estimation framework was proposed in Kay (1989), relying on the time difference of the phase spectrum. The method was shown to provide an accurate frequency estimation of a complex-valued sinusoid in white Gaussian noise. The estimator avoided the phase unwrapping difficulties. In Lagrange and Marchand (2007), and Kay (1989), the authors demonstrated that the phase difference estimator shows the most stable performance in frequency estimation in terms of introducing a low bias and variance. For example, Lagrange and Marchand (2007) proposed an IF estimator using the kth frequency component at frame l and assuming a hop size of S samples between two consecutive frames, as used in (2.29).

It is important to note that the phase-difference-based frequency estimators originally come from the phase vocoder literature (Dolson 1986), where it is assumed that the frequency of one sinusoid remains constant within the time interval of two consecutive short-time segments. The IF principle was used for phase reconstruction across time in Mehmetcik and Ciloglu (2012). The method relies on the fact that the temporal derivative of phase can be approximated in STFT as

$$\Delta\phi(k, l) = \mathrm{ARG}[e^{j(\phi(k,l)-\phi(k,l-1))}]. \tag{2.31}$$

In Boashash (1992b) an overview of different instantaneous frequency estimators is given, while for a detailed review of the concepts of IF, see Boashash (1992a). Finally, the change of the IF from frame to frame and its distribution has been successfully used to detect transient sounds such as onsets in music signals (Bello *et al.* 2005).

2.4.3 Baseband Phase Difference

Consider a signal composed of a single harmonic. Let ω_h be closest to the kth STFT band with center frequency ω_k. Figure 2.5 shows the symbolic spectrum of the signal as its

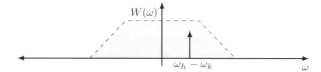

Figure 2.5 The baseband representation of band *k* for a symbolic spectrum composed of one harmonic. The prototype window function spectrum *W*(*ω*) suppresses the impact of the adjacent frequency bands, but not the one closest to the frequency bin of interest, *k* (Krawczyk and Gerkmann 2012).

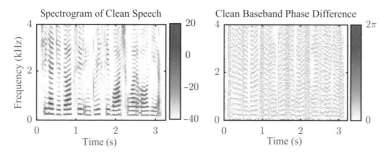

Figure 2.6 Spectrogram in dB (left) and baseband phase difference (BPD; right) calculated for a clean speech signal used in the short-time Fourier transform phase improvement (STFTPI) method (Krawczyk and Gerkmann 2012).

baseband-transformed version in the kth STFT band. Krawczyk and Gerkmann (2012) proposed to reconstruct the STFT phase across frequency, where they transformed the bandpass phase information of STFT into its baseband representation, denoted by $X_\mathrm{B}(k, l)$, by modulating each kth STFT band as follows:

$$X_\mathrm{B}(k, l) = X(k, l)e^{-j\omega_k lS}. \tag{2.32}$$

The baseband phase difference (BPD) representation was defined as (Krawczyk and Gerkmann 2014)

$$\Delta_\mathrm{B}\phi(k, l) = \mathrm{ARG}[\Delta\phi(k, l) - \omega_k S]. \tag{2.33}$$

Figure 2.6 plots the BPD (right) versus the spectrogram (left) for a female utterance, selected from the TIMIT database, saying "she had your dark suit in greasy wash water all year." The BPD representation reveals the temporal and spectral structure of phase that was not visible in the instantaneous phase of STFT. Formant and envelope structure are still not that visible.

2.4.4 Harmonic Phase Decomposition

2.4.4.1 Background on the Harmonic Model

The short-time Fourier transform, although useful, often suffers from the redundancy in representing sparse signals including speech. Its signal independence due to employing fixed basis functions results in a non-instinctual representation of the phase signal spectrum. In contrast, a sinusoidal or harmonic model is often reported to be a popular choice for an efficient representation of a time domain speech signal. It has been successfully used for speech analysis/synthesis (McAulay and Quatieri 1986) and speech coding (Agiomyrgiannakis and Stylianou 2009).

Given the time domain signal $x(n)$, and following segmentation and windowing, for the lth speech segment of length $t(l)$ we obtain

$$x_w(n', l) = x(n' + t(l))w(n'), \tag{2.34}$$

with n' defined as the STFT time index, $n' \in [-(N_w - 1)/2, (N_w - 1)/2)]$ and N_w as the segment length in samples, $w(n')$ and $t(l)$ denoting the analysis window following a pitch-synchronous segmentation $t(l) = t(l-1) + 0.25/f_0(l-1)$ (Degottex and Erro 2014). Using the sinusoidal model, the coded speech for the lth frame denoted by $\hat{x}_w(n, l)$ is represented as sum of a finite number of sinusoids:

$$\hat{x}_w(n', l) \approx \sum_{h=1}^{H_l} a(h, l) \cos(h\omega_0(l)n' + \psi(h, l)) \, w(n'), \tag{2.35}$$

where each hth sinusoid is characterized by the triple parameters consisting of amplitude, harmonic frequency, and phase, denoted by $a(h, l)$, $hf_0(l)$, and $\psi(h, l)$, respectively. Further, we define the model order H_l, which is an integer, often defined as the largest integer yielding $H_l f_0 \leq \frac{f_s}{2}$. Here, we assume that the signal follows a harmonic model where each sinusoid frequency is a multiple of a fundamental frequency f_0 and, hence, we have $f_h(l) = hf_0(l)$.

Apart from the general representation of the harmonic model given in (2.35), several different versions of sinusoidal models have been proposed in the literature. The previous works on sinusoidal modeling were pioneered by McAulay and Quatieri (1986), followed by the sinusoidal model of George and Smith (1997), and the harmonic plus noise model proposed by Stylianou (2001). Most recently, Degottex and Erro (2014) proposed a *harmonic model plus phase distortion* (HMPD), capable of representing both voiced and unvoiced speech with a high quality. In the following, we only focus on HMPD (Degottex and Erro 2014) as it provides a framework to decompose the instantaneous phase at harmonics, providing a more accessible representation than the instantaneous phase.

2.4.4.2 Phase Decomposition using the Harmonic Model

The instantaneous phase at each harmonic h and frame l is decomposed to a linear part and an unwrapped part given by:

$$\psi(h, l) = \underbrace{h \sum_{l'=0}^{l} \omega_0(l') \, (t(l') - t(l' - 1))}_{\text{Linear phase: } \psi_{\text{lin}}(h,l)} + \underbrace{\angle V(h, l) + \psi_d(h, l)}_{\text{Unwrapped phase: } \Psi(h,l)}, \tag{2.36}$$

where the first term explains the cyclic wrapping of the instantaneous phase across time and will be referred to as the *linear phase* in the following. It only depends on the fundamental frequency $f_0(l)$. The second term in (2.36), denoted as $\angle V(h, l)$, is the minimum phase part, known as the phase response of the vocal tract filter V sampled at harmonic h. The minimum phase characteristics originate from the all-pole model assumption.[2]

---

2 Details about speech production and all-pole modeling of the vocal tract system can be found in Chapter 2 of Vary and Martin (2006).

Assuming that the vocal tract filter follows an autoregressive process, it is possible to estimate the minimum phase part from the magnitude of the signal using the Hilbert transform (for more details, see Oppenheim and Schafer 1989).

The last term in (2.36) is called the *phase dispersion* and is denoted by $\psi_d(h, l)$. It is also called the source shape term since it characterizes the pulse shape. It captures the stochastic characteristic of the phase at harmonic h which is not explained by the linear phase or the minimum phase terms. Phase dispersion has been successfully used in harmonic speech coding where vector quantization (Pobloth and Kleijn 2003) or the wrapped Gaussian distribution (Agiomyrgiannakis and Stylianou 2009) were employed for harmonic phase coding. The superposition of the second and the third term constitutes the *unwrapped phase* denoted by $\Psi(h, l)$. As the deterministic contribution of the linear phase is excluded, $\Psi(h, l)$ can be considered as a non-deterministic random variable. This representation for harmonic phase will be discussed in detail in Section 2.4.5.

Having defined the harmonic phase concept, we now proceed with highlighting three recent representations, derived from the harmonic phase, providing useful insights into phase interpretation.

2.4.5 Phasegram: Unwrapped Harmonic Phase

2.4.5.1 Definitions and Background

The harmonic phase decomposition concept, explained above, is one way to get access to the unwrapped phase, avoiding the sophisticated and restricted phase unwrapping algorithms discussed earlier in Section 2.3.3. By removing the linear phase component, an unwrapped harmonic phase is provided, denoted by $\Psi(h, l)$, and we have:

$$\Psi(h, l) = \psi(h, l) - h \sum_{l'=0}^{l} \omega_0(l')(t(l') - t(l' - 1)). \tag{2.37}$$

The unwrapped phase at harmonics is known for its desired smooth changes across time, whereas for non-speech regions it shows no smooth behavior (but rather some randomized pattern). The smoothness property of the unwrapped harmonic phase has been incorporated to develop several interpolation techniques used for harmonic speech synthesis, including cubic interpolation (McAulay and Quatieri 1986), phase-locked pitch-synchronous, and quadratic phase spline (Agiomyrgiannakis 2015). Further, Mowlaee and Kulmer applied the smoothness property of unwrapped harmonic phase to enhance a noisy speech signal by modifying the noisy spectral phase (Mowlaee and Kulmer 2015a,b; Kulmer and Mowlaee 2015a,b; Kulmer *et al.* 2014).[3]

2.4.5.2 Circular Mean and Variance

To capture the uncertainty in the harmonic phase representation, instead of a uniform prior, one could fit a von Mises distribution (Evans *et al.* 2000, p. 191) with the following probability density function:

$$\mathcal{VM}(\mu_c(h, l), \kappa(h, l)) = \frac{e^{\kappa(h,l)\cos(\Psi(h,l)-\mu_c(h,l))}}{2\pi I_0(\kappa(h, l))}, \tag{2.38}$$

---

3 A detailed explanation of these methods and a comparative study will be presented in Chapter 3.

where $\mu_c(h, l)$ and $\kappa(h, l)$ denote the circular mean and the concentration parameters, respectively, and $I_0(\cdot)$ is the modified Bessel function of first kind. The von Mises distribution is known to have the maximum entropy distribution of phase and therefore provides a good fit to model the uncertainty in the estimated harmonic phase.

In the following, we present how to estimate the von Mises parameters, mean and variance. In order to obtain the parameters of the von Mises distribution for the unwrapped phase $\Psi(h, l)$, we first need to consider the sample mean, given by (Kulmer and Mowlaee 2015b):

$$z(h, l) = \frac{1}{|\mathcal{R}|} \sum_{l' \in \mathcal{R}} e^{j\Psi(h, l')}, \tag{2.39}$$

where $z(h, l)$ is the sample mean and $\mathcal{R}$ denotes the set of frames within 20 ms, a good estimate for the short-time stationarity of speech signals. The circular mean and the circular variance are given by:

$$\mu_c(h, l) = \angle z(h, l),$$
$$\sigma_c^2(h, l) = 1 - |z(h, l)|. \tag{2.40}$$

The concentration parameter $\kappa(h, l)$ is calculated by inverting the following formula (Fisher 1995) as proposed in Barens (2009):

$$\sigma_c^2(h, l) = 1 - A(\kappa(h, l)), \tag{2.41}$$

where $A(\kappa(h, l)) = \frac{I_1(\kappa(h,l))}{I_0(\kappa(h,l))}$ is called the ratio of the modified Bessel function of the first kind. For further details with regard to the approximation of the inverse function $A^{-1}(\kappa(h, l))$, see Hill (1981).

It is important to note that $\kappa(h, l) \to 0$ indicates a large variance in phase showing a uniform distribution while $\kappa(h, l) \to \infty$ models the maximum certainty as a Dirac delta. Through a computer experiment, we explore the statistical characteristics of the phase modeled by a von Mises distribution in terms of mean and variance. We simulate 200 000 random angles from a von Mises distribution with zero mean and different values of the concentration parameter as $\kappa = 0.2, 1, 2, 5, 10,$ and 50. The results are shown in Figure 2.7 on the next page. The distribution ranges between a uniform prior (maximum uncertainty) to a Dirac delta (deterministic with maximum certainty) when κ is set to 0 or grows to infinity, respectively.

2.4.6 Relative Phase Shift

Saratxaga *et al.* (2012) proposed a harmonic phase representation called relative phase shift (RPS), where the relation between the phase of the hth harmonic multiple to the phase of the fundamental frequency $h = 1$ is defined as:

$$\text{RPS}(h, l) = \psi(h, l) - h\psi(1, l). \tag{2.42}$$

It is important to note that the RPS representation is independent of $f_0(l)$:

$$\text{RPS}(h, l) = \Psi(h, l) + h \sum_l f_0(l)\Delta t + \angle V(h, l)$$
$$- h\Psi(h, 1) - h \sum_l f_0(l)\Delta t - h\angle V(h, 1)$$
$$= \Psi(h, l) + \angle V(h, l) - h\Psi(h, 1) - h\angle V(h, 1). \tag{2.43}$$

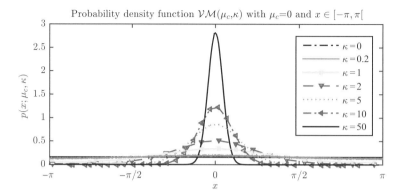

Figure 2.7 Non-uniform distribution for spectral phase in the form of von Mises characterized by mean μ_c and concentration κ, ranging between uniform distribution and Dirac delta.

The RPS is known to be smooth (Saratxaga *et al.* 2012) and hence circumvents the phase wrapping problem without requiring a pitch pulse onset or the GCI estimation. The differences in the initial phase shift of the sinusoids determine the signal waveform. Therefore, the RPS values remain constant when the temporal shape of the signal is constant.

The useful structure in the RPS representation has been successfully used for single-channel speech enhancement via phase modification (Kulmer *et al.* 2014; Mowlaee and Kulmer 2015b), where a frequency smoothing filter was applied on the noisy phase spectrum. Further, derived features from RPS have been successfully used for speaker recognition application (de Leon *et al.* 2011; Sanchez *et al.* 2015), reporting improved performance compared to the conventional MFCC-based systems.

2.4.7 Phase Distortion

Degottex and Erro (2014) proposed the so-called *phase distortion (PD)* to model the randomness in the voice excitation, capturing the noisiness and breathiness of inharmonic parts in speech. PD is defined by removing the contributions of the minimum and linear phase parts from the instantaneous phase given by

$$\mathrm{PD}(h, l) = \psi(h + 1, l) - \psi(h, l) - \psi(1, l). \tag{2.44}$$

The PD representation was further used to model the glottal shape of the signal (Degottex *et al.* 2011, 2012). The finite difference used in PD calculation renders it similar to the group delay. Degottex and Erro (2014) assumed a statistical model for PD characterized by the mean and the variance of a wrapped normal distribution. The phase distortion was modeled by fitting a wrapped Gaussian distribution given by (Degottex and Obin 2014)

$$\mathrm{PD}(h, l) = \mathcal{WG}(\mathrm{PDM}(h, l), \sigma_c(h, l)), \tag{2.45}$$

where $\mathcal{WG}(\cdot)$ denotes the wrapped normal distribution, the *phase distortion mean* (PDM) is

$$\mathrm{PDM}(h, l) = \angle \left(\frac{1}{|\mathcal{R}|} \sum_{l' \in \mathcal{R}} e^{j\mathrm{PD}_{l'}(h, l')} \right), \tag{2.46}$$

and $\sigma_c(h, l)$ is the *phase distortion standard deviation* (PDD) defined at harmonic multiples of $f_0(l)$ and calculated as

$$\sigma_c(h, l) = \sqrt{-2 \ln \left| \frac{1}{|\mathcal{R}|} \sum_{l' \in \mathcal{R}} e^{j \text{PD}_{l'}(h, l')} \right|}, \tag{2.47}$$

where $\mathcal{R}$ defines the set of frames adjacent to the current processing frame indexed l. It was reported in Degottex and Obin (2014) that hoarse or breathy voices show a large PDD. In contrast, a voiced phoneme is represented by a relatively low PDD contribution, in particular below 5 kHz. A creaky voice has close to zero phase distortion mean value.

Figure 2.9 on the next page shows an example demonstrated in Degottex *et al.* (2014) on glottal source representation. The example shows how the phase distortion statistics as mean and standard deviation are useful to classify different voicing states, from left to right: onset, voiced, and offset. The results are shown as (top) time domain, (middle) phase distortion mean, and (bottom) PDD. The onset frame contributes to a large change in PDM and PDD, while a voiced frame is characterized by a close to zero phase distortion.

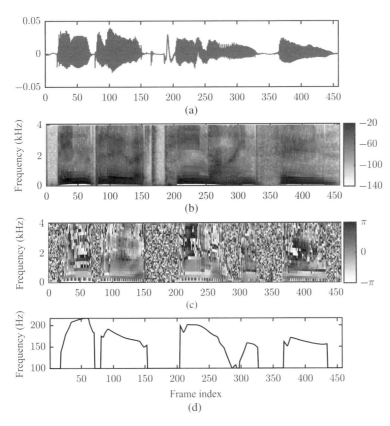

Figure 2.8 (a) Time domain signal for female speech, (b) spectrogram in dB, (c) RPS, (d) fundamental frequency.

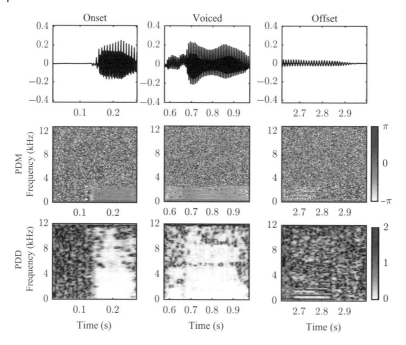

Figure 2.9 Example showing how phase distortion features as mean and deviation are used to classify different voicing states: (left) onset, (middle) voiced, (right) offset. The results are shown as (top) time domain, (middle) phase distortion mean (PDM), and (bottom) phase distortion standard deviation (PDD).

Finally, it is important to emphasize that the last two representations, RPS and PD are related by the following equation (Degottex and Erro 2014):

$$PD(h, l) = \widetilde{RPS}(h + 1, l) - \widetilde{RPS}(h, l), \tag{2.48}$$

where $\widetilde{RPS}(h, l)$ is the RPS representation after removing the minimum phase frequency response corresponding to the amplitude envelope and is defined as

$$\widetilde{RPS}(h, l) = \psi_d(h, l) - h\psi_d(1, l), \tag{2.49}$$

this finally results in

$$PD(h, l) = \tilde{\psi}(h + 1, l) - \tilde{\psi}(h, l) - \tilde{\psi}(1, l), \tag{2.50}$$

defining $\tilde{\psi}(h, l)$ as the instantaneous phase where the minimum phase frequency response corresponding to the amplitude envelope has been removed, given by:

$$\tilde{\psi}(h, l) = \psi(h, l) - \angle V(hf_0(l), l). \tag{2.51}$$

It is important to note that the RPS depends on the harmonic number h explicitly, which is circumvented in PD. Figure 2.8 shows the harmonic structure in RPS for a female utterance selected from the GRID corpus (Cooke *et al.* 2006). The values in RPS stay almost constant for the regions where the temporal shape of the signal is not varying (e.g., for strong voiced frames).

2.5 Experiments

In this section we present the following experiments to explain how the concepts discussed are used in practice:

- Experiment 2.1: One-dimensional phase unwrapping
- Experiment 2.2: Comparative study of phase unwrapping methods
- Experiment 2.3: Comparative study of group delay spectra
- Experiment 2.4: Circular statistics of the harmonic phase
- Experiment 2.5: Circular statistics of the spectral phase
- Experiment 2.6: Comparative study of phase representations.

2.5.1 Experiment 2.1: One-Dimensional Phase Unwrapping

Here, we demonstrate the one-dimensional phase unwrapping problem. Through this experiment the phase unwrapping process and its robustness to noise will be demonstrated. We consider a sine waveform as shown in Figure 2.10 panel (a). Panel (b) shows the wrapped phase signal where 2π jumps occur as the original continuous phase range exceeds the valid range of $[-\pi, \pi[$. Phase unwrapping is required to remove the phase jump and to return to the continuous phase signal shown in panel (f).

2.5.1.1 Clean Signal Scenario

The original wrapped phase is shown in (b), which consists of four discontinuities. The procedure for phase unwrapping to remove each wrap from left to right, using the *discontinuity detection* (DD) method (Oppenheim and Schafer 1989), is shown throughout plots (b) to (f). The idea is to calculate the difference between the current sample and its adjacent left-hand neighbor to detect any discontinuity. If this difference is larger than π, 2π is subtracted from this sample and also from all samples to the right of it. Otherwise, if the difference between the two samples is smaller than $-\pi$, 2π is added to this sample as well as to all the samples to its right. The unwrapped signal after applying the

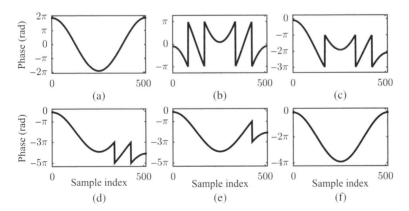

Figure 2.10 Example inspired by Gdeisat and Lilley (2011) to show the process of phase unwrapping using the DD method applied to a cosine waveform; starting from the wrapped phase (b), via adding/subtracting 2π jumps to remove the wraps, sequentially shown in (c)–(f), for the four wraps in the wrapped phase signal shown in (b).

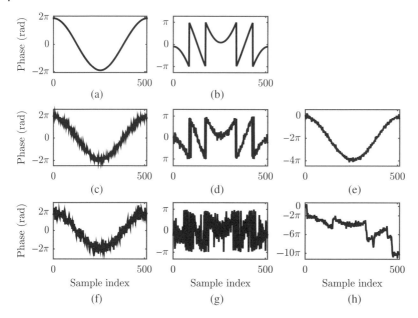

Figure 2.11 Example inspired by Gdeisat and Lilley (2011) to show the process of phase unwrapping using the DD method applied to a cosine waveform corrupted with additive noise. One-dimensional phase unwrapping problem: top panel: (a) continuous phase, (b) wrapped phase, (c) mild noisy version, (d) wrapped phase, (e) unwrapped phase for mild noise, (f) intense noisy version, (g) wrapped phase, (h) unwrapped phase for intense noise.

procedure for the four wraps is shown in panel (f). The MATLAB® implementation for this experiment can be found in *Exp2_1a.m* in the *PhaseLab Toolbox*, described in the appendix.

2.5.1.2 Noisy Signal Scenario

The aforementioned phase unwrapping procedure works properly when the underlying signal is not corrupted with noise. In the following, we repeat the experiment for a noisy scenario. We consider two noise scenarios: a mild noise variance and an intense noise variance. The results are shown in Figure 2.11. The fake wrap in the intense noise scenario (see panel (h)) will affect a significant part of the signal, as the error propagates to the rest of the samples to the right of the fake detected wrap. In fact, for a noisy scenario, there will be an ambiguity in the phase wraps as it is not known if they are true or wrongly produced by the noise contribution. This example explains the reason why phase unwrapping is a challenging task, as one single error in the phase wrap detection affects the subsequent samples of the signal due to the propagation of errors. The MATLAB® implementation for this experiment can be found in *Exp2_1b.m* in the *PhaseLab Toolbox* described in the appendix.

2.5.2 Experiment 2.2: Comparative Study of Phase Unwrapping Methods

Here, we present a comparative study conducted in Drugman and Stylianou (2015) between different phase unwrapping methods studied in this chapter (for the list of the methods see Table 2.1). The results are shown for a signal of length N parameterized as $N = 10^{\alpha}$, with α ranging from 1 to 3.4 by steps of 0.2. This parametrization helps to address the impact of the data length on the performance of the phase unwrapping methods. Real speech signals were taken from the ARCTIC database (Black 2014). The following unwrapping methods are considered in the performance evaluation:

1) **DD ($N_{DFT} = 64$):** DD method (Oppenheim and Schafer 1989) with $N_{DFT} = 64$ DFT points.
2) **DD ($N_{DFT} = 1024$):** DD method with $N_{DFT} = 1024$.
3) **Match (M):** Matched method that iteratively increases (doubles) the number of DFT points (N_{DFT}) until two consecutive unwrapped phase estimates match.
4) **PF (using roots):** PF method relying on the calculation of the zeros of the polynomial, computed by the eigenvalues of the companion matrix (Edelman and Murakami 1995) (*roots* command in MATLAB®).
5) **PF (using lroots):** PF method using the FFT-based search grid (Sitton *et al.* 2003) (*lroots* command in MATLAB®).
6) **D:** The method proposed by Drugman and Stylianou (2015).[4]

The phase unwrapping algorithms were compared in terms of two criteria:

- the resulting error rate defined as the proportion of erroneous estimates,
- computation time.

As the ground truth for calculating the accuracy, the unwrapped phase calculated by PF (using *lroots*) is used as the reference.

The results are shown in Figure 2.12 in terms of computation time (top) and error rate (bottom). The method proposed by Drugman and Stylianou (2015) (marked as D) presents a perfect unwrapping. The matching method (M) performs worst and the DD method (DD) is prone to errors. The error rate produced by DD increases with N and can be reduced by increasing the number of DFT points (compare the $N_{DFT} = 64$ and $N_{DFT} = 1024$ settings).

Apart from the resulting error rate introduced by the selected phase unwrapping method, it is also relevant to study the complexity via the computational time to get an estimate on how long it takes. A comparative study on the computation time required for each phase unwrapping method is shown in the upper panel of Figure 2.12. The MATLAB® implementations were simulated on an Intel Core i7 3.0 GHz CPU with 16 GB of RAM (Drugman and Stylianou 2015). In terms of computational complexity, the two PF-based methods require $O(N^3)$ and $O(N^2)$ orders of complexity, while the FFT-based method requires only $O(N \log(N))$. In contrast, method D (Drugman and Stylianou 2015) balances a good trade-off between low computational load and perfect phase unwrapping accuracy.

2.5.3 Experiment 2.3: Comparative Study on Group Delay Spectra

Here, we present a comparison of the group delay spectra provided by the different extensions explained in Section 2.4.1 (for a summary, see Table 2.3). In the experiment, a voiced frame of a male utterance saying "lay red by l one soon" is selected with a frame setup of 50% overlap, a window length of 30 ms and a sampling frequency of 25 kHz.

---

4 The function *drugmanunwrapping.m* is part of the COVAREP toolbox (Degottex *et al.* 2014) available at http://covarep.github.io/covarep.

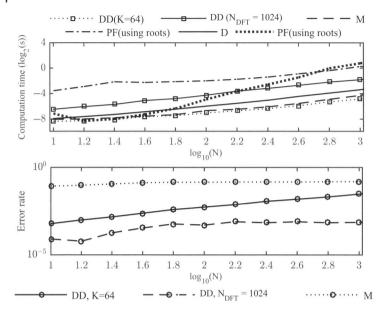

Figure 2.12 Computation time (top) and error rate results (bottom) for the phase unwrapping methods for speech signals as a function of *N* for different phase unwrapping methods.

Figure 2.13 shows demonstrations of the group delay representations listed in Table 2.3. The MATLAB® implementation for this experiment is found in *Exp2_3.m* in the *Phase-Lab Toolbox* described in the appendix.

The following observations can be made:

- Both LPGD and CGD representations reveal more information regarding the spectral peaks and valleys in the group delay spectrum compared to when the Fourier transform is used directly.
- Comparison of the spectral peaks and valleys between group delay representations reveals that MGD (Hegde *et al.* 2007) or LPGD (Rajan *et al.* 2013) have sharper peaks, and hence are more robust to noise compared to the others.
- The formants in the modified group delay spectra are presented with a higher frequency resolution, resulting in a robust presentation against noise. This robustness is an attractive property that has been reported quite helpful in speech processing applications, for example automatic speech and speaker recognition (for some examples, see Section 2.4.1).

2.5.4   Experiment 2.4: Circular Statistics of the Harmonic Phase

Here, we show how to apply circular statistics to real speech signals. In particular, we show how phase statistics follow a von Mises distribution for speech signals. In our experiments we show the circular mean and variance for the first three harmonics of voiced/unvoiced speech frames for a clean speech scenario.

Figure 2.14 shows the results. The unwrapped phase is obtained using (2.37), and the circular statistics parameters mean and variance are calculated using (2.40) considering frames $\mathcal{R}$ within 20 ms. The top panel shows a female utterance taken from the

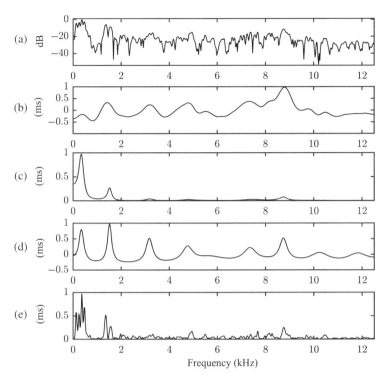

Figure 2.13 Group delay representations listed in Table 2.3 shown for a voiced speech segment: (a) FFT log-magnitude, (b) modified group delay (MGD; Hegde *et al.* 2007), (c) LPC, (d) CGD (Bozkurt *et al.* 2007), and (e) LPGD (Rajan *et al.* 2013).

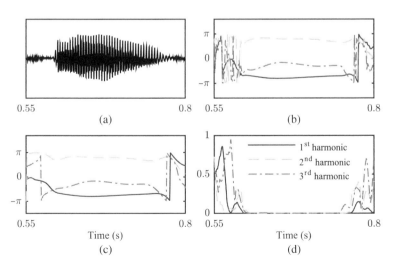

Figure 2.14 (a) Time domain representation for clean speech; (b) unwrapped phase shown for the first three harmonics, (c) mean, and (d) circular variance.

GRID corpus. The result of the unwrapped phase, circular mean, and phase variance are shown for the first three harmonics in panels (b)–(d). The results are shown for a clean scenario. It can be seen that at harmonics the desired clean phase preserves a smooth characteristic across frames. This corroborates with the high voice quality for small phase variance as reported in Koutsogiannaki *et al.* (2014).

2.5.5 Experiment 2.5: Circular Statistics of the Spectral Phase

The previous experiment illustrated circular statistic parameters of a real speech signal evaluated at the first three harmonics. It can be concluded that the circular variance parameter reveals lot of information regarding the speech production process. To get even more insight, we extend Experiment 2.4 by applying circular statistics at frequency bins. The focus is set on the variations of the fundamental frequency that violate the assumption of stationarity and lead to a biased parameter estimation. We introduce a modified STFT with a constant frame shift and frame lengths of $N_w(l)$ dependent on fundamental frequency $f_0(l)$ for each frame l. Given an accurate $f_0(l)$, the frame lengths are calculated as follows:

$$N_w(l) = \left\lfloor m \frac{f_s}{f_0(l)} \right\rfloor, \tag{2.52}$$

with m denoting the desired number of periods per $N_w(l)$, f_s the sampling frequency, and $\lfloor . \rfloor$ the floor function. In the case of unvoiced phonemes, $f_0(l)$ is interpolated between the nearest fundamental frequency estimates. A symmetric, zero-phase window function[5] with length $N_w(l)$ is applied for each frame, yielding the DTFT signal $X_w(e^{j\omega}, l)$. To obtain the DFT $X_w(k, l)$, each frame of the DTFT is sampled with $k = \omega \frac{N_w(l)}{2\pi}$ with $k \in [0, N_w(l) - 1]$. This leads to an $f_0(l)$-related spectrum. The harmonic h of each frame is now located in the $m \cdot h$ frequency bin, independent of the fundamental frequency of frame l. The circular mean $\mu_c(k, l)$ and variance $\sigma_c^2(k, l)$ are calculated for each frequency bin k according to (2.39) and (2.40). Finally, the circular parameters of each frame are interpolated to a frequency scale from 0 to $f_s/2$.

Figure 2.15 illustrates the circular variance of the sentence "The sky that morning was clear and bright blue" for clean speech (left) and additive, white Gaussian noise of

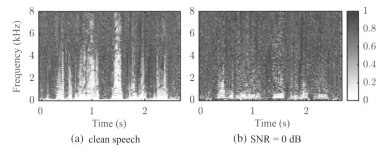

(a) clean speech   (b) SNR = 0 dB

Figure 2.15 Phase variance presentation for (a) clean speech and (b) speech deteriorated by additive, white Gaussian noise. In highly voiced regions, e.g. at time 1 s, the harmonic structure is visible due to their low phase variance. Additive noise increases the phase variance.

---

5 See Chapter 3 for further details.

0 dB (right). We choose a small frame shift of 3 ms to guarantee an accurate parameter estimation. The number of periods per frame length is set to $m = 3$ and a Hamming window is used. In voiced regions the phase variance at harmonics approaches low values. The regions between the harmonics are dominated by noise, which results in an increased phase variance. The implementation used for this experiment is found in *Exp2_5.m* in the *PhaseLab Toolbox* (see the appendix).

2.5.6 Experiment 2.6: Comparative Study of Phase Representations

We show the phase representations and their useful properties explained throughout this chapter (see Table 2.2 for the list of phase representations). The example visualizes the structure in phase for two scenarios: (i) a clean speech signal, and (ii) noisy speech, where clean speech is corrupted with noise. In this experiment, the robustness of the representations and their patterns will be commented on and graphically analyzed. To produce the results explained in this experiment, we used the implementation in file *Exp2_6.m* in the *PhaseLab Toolbox*.

Figure 2.16 shows various phase representations described in this chapter. The results are shown for a female utterance selected from the GRID corpus (Cooke *et al.* 2006).[6] The utterance is "bin blue at o seven now," which consists of different phoneme classes such as plosives [b], vowels [U:], and fricatives [s]. The results are shown for clean speech (left) and noisy speech (right). To produce the noisy signal, the clean speech was mixed with white noise mixed at a global SNR of 0 dB. Figure 2.16 shows, from top to bottom: (a) spectrogram, (b) instantaneous phase, (c) group delay, (d) instantaneous frequency, (e) unwrapped harmonic phase (phasegram), (f) phase distortion standard deviation (PDD), and (g) relative phase shift (RPS). Note, the PDD and RPS are defined for harmonics h rather than DFT bins k, and for the comparison they are interpolated to frequency bins k. Several speech characteristics can be inferred from the clean speech representations:

- The instantaneous phase obtained from the Fourier spectral analysis presents no harmonic structure and yields a random pattern following the uniform distribution of the DFT coefficients.
- For clean speech signals, both phase variance and group delay plots follow the harmonic structure of speech visible in the spectrogram.
- The PDD shows a binary structure whereby its contribution varies from a close to zero pattern at low frequencies to a randomized pattern for high frequencies. The randomized pattern of PDD observed at high frequencies explains the high variance in voice excitation in the speech production.
- The RPS representation uncovers the harmonic structure of the phase spectrum across frequency, which was not observable in the instantaneous phase. This is made clear by comparing the patterns displayed in the last row in Figure 2.16.

On the right column in Figure 2.16, the same representations are given for noisy speech. The harmonic structure in the phase representations is mostly lost due to the detrimental contribution of the additive noise. In addition, we see that:

---

6 The utterances in GRID are structured as command-like and consist of six distinguished units consisting of command, color, preposition, letter, digit, and adverb.

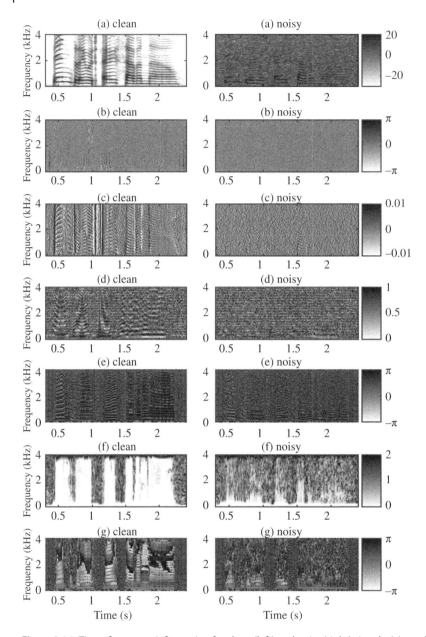

Figure 2.16 Time–frequency information for clean (left) and noisy (right) signals: (a) amplitude spectrogram in dB, (b) instantaneous phase, (c) group delay, (d) instantaneous frequency (IF), (e) phasegram, (f) phase distortion deviation (PDD), and (g) relative phase shift (RPS).

- Higher-order harmonics that are often inherent at lower amplitudes are easily masked, and hence lost in the spectrogram. Similarly, the phasegram and group delay plots confirm the loss of harmonic structure which is useful in speech signal processing.
- The phase variance is increased at harmonics, leading to perceptual degradation and buzzyness in the reconstructed speech signal. In particular, Koutsogiannaki *et al.* (2014) reported that the phase variance is a reliable measure for assessing voice quality. Therefore, the clear difference in the patterns of phase variance for clean and noisy speech quantitatively justify how much the voice quality is degraded. With the same motivation, the circular variance has been used to enhance the noisy spectral phase for improved signal reconstruction in speech enhancement (Mowlaee and Kulmer 2015a,b; Kulmer and Mowlaee 2015a,b; Maly and Mowlaee 2016).

2.6 Summary

This chapter has provided an overview of phase signal processing that will be a useful asset for the later chapters of the book. Starting from the short-time Fourier spectrum and its resulting instantaneous phase, we highlighted the main difficulties in phase processing due to its wrapping issue. Several methods to obtain an unwrapped phase were discussed. A comparative study between the phase unwrapping methods in terms of their accuracy and computational cost was demonstrated. Throughout harmonic phase decomposition, the details with regard to a non-uniform distribution assumption for harmonic phase were established. Such a non-uniform distribution of phase is capable of explaining the uncertainty of phase as a random variable around a mean value characterized by a von Mises distribution.

We further introduced several useful phase representations other than the STFT instantaneous phase, revealing the harmonic structure of the instantaneous phase. Some of these representations rely on the temporal or frequency derivative, called instantaneous frequency and group delay, while others rely on the harmonic phase decomposition principle, for example phase variance, relative phase shift, phase distortion, and unwrapped harmonic phase. Finally, through numerous examples and experiments, we showed how the new phase representations could be used as valuable tools to obtain useful information about the phase spectrum of speech. A comparative study between the phase-based features for clean and noisy speech has been presented.

References

Y. Agiomyrgiannakis, *VOCAINE: The Vocoder and Applications in Speech Synthesis*, Proceedings of the IEEE International Conference on Acoustics, Speech and Signal Processing (ICASSP), pp. 4230–4234, 2015.

Y. Agiomyrgiannakis and Y. Stylianou, Wrapped Gaussian mixture models for modeling and high-rate quantization of phase data of speech, *IEEE Transactions on Audio, Speech, and Language Processing*, vol. 17, no. 4, pp. 775–786, 2009.

L. D. Alsteris and K. K. Paliwal, Short-time phase spectrum in speech processing: A review and some experimental results, *Signal Processing*, vol. 17, no. 3, pp. 578–616, 2007.

P. Barens, CircStat: A MATLAB toolbox for circular statistics, *Journal of Statistical Software*, vol 31, no. 10, pp. 1–21, 2009.

J. Bello, L. Daudet, S. Abdallah, C. Duxbury, and M. Davis, A tutorial on onset detection in music signals, *IEEE Transactions on Speech and Audio Processing*, vol. 13, no. 5, pp. 1035–1047, 2005.

A. W. Black, *CMU ARCTIC speech synthesis database*, online, www.festvox.org/cmu_arctic, 2014.

B. Boashash, Estimating and interpreting the instantaneous frequency of a signal. I. Fundamentals, *Proceedings of the IEEE*, vol. 80, no. 4, pp. 520–538, 1992a.

B. Boashash, Estimating and interpreting the instantaneous frequency of a signal. II. Algorithms and applications, *Proceedings of the IEEE*, vol. 80, no. 4, pp. 540–568, 1992b.

F. Bonzanigo, An improvement of Tribolet's phase unwrapping algorithm, *IEEE Transactions on Acoustics, Speech and Signal Processing*, vol. 26, no. 1, 1978.

B. Bozkurt, *Zeros of the z-transform (ZZT) Representation and Chirp Group Delay Processing for the Analysis of Source and Filter Characteristics of Speech Signals*, PhD Thesis, Faculty Polytechnique De Mons, Belgium, 2005.

B. Bozkurt and L. Couvreur, *On the Use of Phase Information for Speech Recognition*, Proc. 13th European Signal Processing Conference (EUSIPCO), 2005.

B. Bozkurt, L. Couvreur, and T. Dutoit, Chirp group delay analysis of speech signals, *Speech Communication*, vol. 49, no. 3, pp. 159–176, 2007.

J. R. Carson and T. C. Fry, Variable frequency electric circuit theory with application to the theory of frequency modulation, *Bell System Technical Journal*, vol. 16, pp. 513–540, 1937.

L. Cohen, Time-frequency distributions: A Review, *Proceedings of the IEEE*, vol. 77, no. 7, pp. 941–981, 1989.

M. Cooke, J. Barker, S. Cunningham, and X. Shao, An audio-visual corpus for speech perception and automatic speech recognition, *The Journal of the Acoustical Society of America*, vol. 120, pp. 2421–2424, 2006.

G. Degottex, A. Roebel, and X. Rodet, *Function of Phase-Distortion for Glottal Model Estimation*, Proceedings of the IEEE International Conference on Acoustics, Speech and Signal Processing (ICASSP), pp. 4608–4611, 2011.

G. Degottex, E. Godoy, and Y. Stylianou, *Identifying Tenseness of Lombard Speech Using Phase Distortion*, Proc. The Listening Talker: An Interdisciplinary Workshop On Natural and Synthetic Modification of Speech, LISTA Workshop, 2012.

G. Degottex and N. Obin, *Phase Distortion Statistics as a Representation of the Glottal Source: Application to the Classification of Voice Qualities*, Proc. International Conference on Spoken Language Processing (INTERSPEECH), 2014.

G. Degottex and D. Erro, A uniform phase representation for the harmonic model in speech synthesis applications, *EURASIP Journal on Audio, Speech, and Music Processing*, pp. 1–16, 2014.

G. Degottex, J. Kane, T. Drugman, and S. Scherer, *COVAREP: A Collaborative Voice Analysis Repository for Speech Technologies*, Proceedings of the IEEE International Conference on Acoustics, Speech and Signal Processing (ICASSP), pp. 960–964, 2014.

M. Dolson, The phase vocoder: A tutorial, *Computer Music Journal*, vol. 10, no. 4, pp. 14–27, 1986.

T. Drugman and Y. Stylianou, *Fast and Accurate Phase Unwrapping*, Proceedings of the International Conference on Spoken Language Processing (INTERSPEECH), pp. 1171–1175, 2015.

A. Edelman and Y. Murakami, Polynomial roots from companion matrix eigenvalues, *Mathematics of Computation*, vol. 64, pp. 763–776, 1995.

M. Evans, N. Hastings, and B. Peacock, *Statistical Distributions*, pp. 189–191, Wiley & Sons, 2000.

N. I. Fisher, *Statistical Analysis of Circular Data*, Cambridge University Press, 1995.

D. Friedman, *Instantaneous-Frequency Distribution vs. Time: An Interpretation of the Phase Structure of Speech*, Proceedings of the IEEE International Conference on Acoustics, Speech and Signal Processing (ICASSP), pp. 1121–1124, 1985.

J. S. Garofolo, L. F. Lamel, W. M. Fisher, J. G. Fiscus, D. S. Pallett, and N. L. Dahlgren, *DARPA TIMIT Acoustic Phonetic Continuous Speech Corpus CDROM*, NIST, 1993.

M. Gdeisat and F. Lilley, *One-Dimensional Phase Unwrapping Problem*, Course Material, GERI, Liverpool John Moore University, 2011.

E. B. George and M. J. T. Smith, Speech analysis/synthesis and modification using an analysis-by-synthesis/overlap-add sinusoidal model, *IEEE Transactions on Speech and Audio Processing*, vol. 5, no. 5, pp. 389–406, 1997.

D. C. Ghiglia and M. D. Pritt, *Two-Dimensional Phase Unwrapping: Theory, Algorithms, and Software*, John Wiley & Sons, 1998.

G. Goertzel, An algorithm for the evaluation of finite trigonometric series, *The American Mathematical Monthly*, vol. 65, pp. 34–35, 1958.

R. M. Goldstein, H. A. Zebker, and C. L. Werner, Satellite radar interferometry: Two-dimensional phase unwrapping, *Radio Science*, vol. 23, pp. 713–720, 1988.

R. M. Hegde, H. A. Murthy, and G. V. R. Rao, *Application of the Modified Group Delay Function to Speaker Identification and Discrimination*, Proceedings of the IEEE International Conference on Acoustics, Speech and Signal Processing (ICASSP), pp. 517–520, 2004.

R. M. Hegde, H. A. Murthy, and V. R. R. Gadde, Significance of the modified group delay feature in speech recognition, *IEEE Transactions on Audio, Speech and Language Processing*, vol. 15, no. 1, pp. 190–202, 2007.

B. Hideki, L. Jinlin, S. Nakamura, K. Shikano, and H. Kawahara, *Efficient Representation of Short-Time Phase Based on Group Delay*, Proceedings of the IEEE International Conference on Acoustics, Speech and Signal Processing (ICASSP), vol. 2, pp. 861–864, 1998.

G. W. Hill, Evaluation and inversion of the ratios of modified Bessel functions $I_1(x)/I_0(x)$ and $I_{1.5}(x)/I_{0.5}(x)$, *ACM Transactions on Mathematical Software (TOMS)*, vol. 7, no. 2, pp. 199–208, 1981.

V. Kandia and Y. Stylianou, Detection of clicks based on group delay, Special Issue on Marine Mammals Detection and Classification, *Canadian Acoustics*, vol. 36, no. 1, pp. 48–54, 2008.

Z. N. Karam, *Computation of the One-Dimensional Unwrapped Phase*, Masters Thesis, Massachusetts Institute of Technology, 2006.

Z. N. Karam and A. V. Oppenheim, *Computation of the One-Dimensional Unwrapped Phase*, Proceedings of the International Conference on Digital Signal Processing (DSP), pp. 304–307, 2007.

H. Kawahara and I. Masuda, Speech Representation and Transformation based on Adaptive Time-Frequency Interpolation, *Technical Report of IEICE*, EA96-28, pp. 9–16, 1996.

S. Kay, A fast and accurate single frequency estimator, *IEEE Transactions on Acoustics, Speech and Signal Processing*, vol. 37, no. 12, pp. 1987–1990, 1989.

L. H. Keel and S. P. Bhattacharyya, Root counting, phase unwrapping, stability and stabilization of discrete time systems, *Linear Algebra and its Applications*, vol. 351–352, pp. 501–518, 2002.

M. Koutsogiannaki, O. Simantiraki, G. Degottex, and Y. Stylianou, *The Importance of Phase on Voice Quality Assessment*, Proc. International Conference on Spoken Language Processing (INTERSPEECH), 2014.

M. Krawczyk and T. Gerkmann, *STFT Phase Improvement for Single Channel Speech Enhancement*, Proceedings of The International Workshop on Acoustic Signal Enhancement (IWAENC), pp. 1–4, 2012.

M. Krawczyk and T. Gerkmann, STFT phase reconstruction in voiced speech for an improved single-channel speech enhancement, *IEEE/ACM Transactions on Audio, Speech, and Language Processing*, vol. 22, no. 12, pp. 1931–1940, 2014.

J. Kulmer, P. Mowlaee, and M. Watanabe, *A Probabilistic Approach For Phase Estimation in Single-Channel Speech Enhancement Using von Mises Phase Priors*, Proc. IEEE Workshop on Machine Learning for Signal Processing, 2014.

J. Kulmer and P. Mowlaee, Phase estimation in single channel speech enhancement using phase decomposition, *IEEE Signal Processing Letters*, vol. 22, no. 5, pp. 598–602, 2015a.

J. Kulmer and P. Mowlaee, *Harmonic Phase Estimation in Single-Channel Speech Enhancement Using von Mises Distribution and Prior SNR*, Proceedings of the IEEE International Conference on Acoustics, Speech and Signal Processing (ICASSP), pp. 5063–5067, 2015b.

M. Lagrange and S. Marchand, Estimating the instantaneous frequency of sinusoidal components using phase-based methods, *The Journal of the Audio Engineering Society*, vol. 55, no. 5, pp. 385–399, 2007.

P. L. de Leon, I. Hernaez, I. Saratxaga, M. Pucher, and J. Yamagishi, *Detection of Synthetic Speech for the Problem of Imposture*, Proc. IEEE International Conference on Acoustics, Speech and Signal Processing (ICASSP), pp. 4844–4847, 2011.

D. G. Long, *An Exact Numerical Algorithm for Computing the Unwrapped Phase of a Finite-Length Sequence*, Proceedings of the International Conference on Spoken Language Processing (INTERSPEECH), pp. 1782–1785, 1988.

E. Loweimi, S. M. Ahadi, and T. Drugman, *A New Phase-based Feature Representation for Robust Speech Recognition*, Proceedings of the IEEE International Conference on Acoustics, Speech and Signal Processing (ICASSP), pp. 7155–7159, 2013.

E. Loweimi, J. Barker, and T. Hain, *Source-Filter Separation of Speech Signal in the Phase Domain*, Proceedings of the International Conference on Spoken Language Processing (INTERSPEECH), pp. 598–602, 2015.

R. McAulay and T. F. Quatieri, Speech analysis/synthesis based on a sinusoidal representation, *IEEE Transactions on Acoustics, Speech and Signal Processing*, vol. 34, no. 4, pp. 744–754, 1986.

I. McCowan, D. Dean, M. McLaren, R. Vogt, and S. Sridharan, The delta-phase spectrum with application to voice activity detection and speaker recognition, *IEEE Transactions on Audio, Speech, and Language Processing*, vol. 19, no. 7, pp. 2026–2038, 2011.

R. McGowan and R. Kuc, A direct relation between a signal time series and its unwrapped phase, *IEEE Transactions on Acoustics, Speech and Signal Processing*, vol. 30, no. 5, pp. 719–726, 1982.

A. Maly and P. Mowlaee, *On the Importance of Harmonic Phase Modification for Improved Speech Signal Reconstruction*, Proc. IEEE International Conference on Acoustics, Speech and Signal Processing (ICASSP), pp. 584–588, 2016.

E. Mehmetcik and T. Ciloglu, Speech enhancement by maintaining phase continuity between consecutive analysis frames, *The Journal of the Acoustical Society of America*, vol. 132, no. 3, 2012.

P. Mowlaee and J. Kulmer, Phase estimation in single-channel speech enhancement: Limits-potential, *IEEE/ACM Transactions on Audio, Speech, and Language Processing*, vol. 23, no. 8, pp. 1283–1294, 2015a.

P. Mowlaee and J. Kulmer, Harmonic phase estimation in single-channel speech enhancement using phase decomposition and SNR information, *IEEE/ACM Transactions on Audio, Speech, and Language Processing*, vol. 23, no. 9, pp. 1521–1532, 2015b.

P. Mowlaee and R. Saeidi, *Time-Frequency Constraint for Phase Estimation in Single-Channel Speech Enhancement*, Proceedings of The International Workshop on Acoustic Signal Enhancement (IWAENC), pp. 338–342, 2014.

P. Mowlaee, R. Saeidi, and R. Martin, *Phase Estimation for Signal Reconstruction in Single-Channel Speech Separation*, Proc. International Conference on Spoken Language Processing (INTERSPEECH), 2012.

H. A. Murthy and B. Yegnanarayana, Formant extraction from group delay function, *Speech Communication*, vol. 10, no. 3, pp. 209–221, 1991a.

H. Murthy and B. Yegnanarayana, Speech processing using group delay functions, *Signal Processing*, vol. 22, no. 3, pp. 259–267, 1991b.

H. Al-Nashi, Phase unwrapping of digital signals, *IEEE Transactions on Acoustics, Speech, and Signal Processing*, vol. 37, no. 11, pp. 1693–1702, 1989.

A. V. Oppenheim and J. S. Lim, The importance of phase in signals, *Proceedings of the IEEE*, vol. 69, No. 5, pp. 529–541, 1981.

A. V. Oppenheim and R. W. Schafer, *Discrete-Time Signal Processing*, Prentice Hall, 1989.

S. Parthasarathi, R. Padmanabhan, and H. Murthy, Robustness of group delay representations for noisy speech signals, *International Journal of Speech Technology*, vol. 14, no. 4, pp. 361–368, 2011.

H. Pobloth and W. B. Kleijn, Squared error as a measure of perceived phase distortion, *The Journal of Acoustic Society of America*, vol. 114, no. 2, pp. 1081–1094, 2003.

V. K. Prasad, T Nagarajan, and H. A. Murthy, Automatic segmentation of continuous speech using minimum phase group delay functions, *Speech Communication*, vol. 42, no. 3–4. pp. 429–446, 2004.

T. F. Quatieri, *Phase Estimation with Application to Speech Analysis-Synthesis*, Technical Report, Massachusetts Institute of Technology, 1979.

T. F. Quatieri, *Discrete-Time Speech Signal Processing: Principles and Practice*, Prentice Hall, 2001.

T. F. Quatieri and A. V. Oppenheim, Iterative techniques for minimum phase signal reconstruction from phase or magnitude, *IEEE Transactions on Acoustics, Speech and Signal Processing*, vol. 29, no. 6, pp. 1187–1193, 1981.

L. Rabiner, R. W. Schafer, and C. M. Rader, The chirp z-transform algorithm, *IEEE Transactions on Audio and Electroacoustics*, vol. 17, no. 2, pp. 86–92, 1969.

P. Rajan, T. Kinnunen, C. Hanilci, J. Pohjalainen, and P. Alku, *Using Group Delay Functions from All-Pole Models for Speaker Recognition*, Proc. International Conference on Spoken Language Processing (INTERSPEECH), 2013.

J. Sanchez, I. Saratxaga, I. Hernaez, E. Navas, D. Erro, and T. Raitio, Toward a universal synthetic speech spoofing detection using phase information, *IEEE Transactions on Information Forensics and Security*, vol. 10, no. 4, pp. 810–820, 2015.

I. Saratxaga, D. Erro, I. Hernández, I. Sainz, and E. Navas, *Use of Harmonic Phase Information for Polarity Detection in Speech Signals*, Proc. International Conference on Spoken Language Processing (INTERSPEECH), 2009.

I. Saratxaga, I. Hernaez, M. Pucher, and I. Sainz, *Perceptual Importance of the Phase Related Information in Speech*, Proc. International Conference on Spoken Language Processing (INTERSPEECH), 2012.

G. Sitton, C. Burrus, J. Fox, and S. Treitel, Factoring very-high-degree polynomials, *IEEE Signal Processing Magazine*, vol. 20, no. 6, pp. 27–42, 2003.

A. P. Stark and K. K. Paliwal, *Speech Analysis using Instantaneous Frequency Deviation*, Proceedings of the International Conference on Spoken Language Processing (INTERSPEECH), pp. 22–26, 2008.

A. P. Stark and K. K. Paliwal, *Group-Delay-Deviation based Spectral Analysis of Speech*, Proceedings of the International Conference on Spoken Language Processing (INTERSPEECH), pp. 1083–1086, 2009.

K. Steiglitz and B. Dickinson, *Computation of the Complex Cepstrum by Factorization of the z-Transform*, Proceedings of the IEEE International Conference on Acoustics, Speech and Signal Processing (ICASSP), pp. 723–726, 1977.

K. Steiglitz and B. Dickinson, Phase unwrapping by factorization, *IEEE Transactions on Acoustics, Speech and Signal Processing*, vol. 30, no. 6, pp. 984–991, 1982.

Y. Stylianou, Applying the harmonic plus noise model in concatenative speech synthesis, *IEEE Transactions on Speech and Audio Processing*, vol. 9, no. 1, pp. 21–29, 2001.

J. Tribolet, A new phase unwrapping algorithm, *IEEE Transactions on Acoustics, Speech and Signal Processing*, vol. 25, no. 2, pp. 170–177, 1977.

P. Vary and R. Martin, *Digital Speech Transmission: Enhancement, Coding And Error Concealment*, John Wiley & Sons, 2006.

B. Yegnanarayana, D. Saikia, and T. Krishnan, Significance of group delay functions in signal reconstruction from spectral magnitude or phase, *IEEE Transactions on Acoustics, Speech and Signal Processing*, vol. 32, no. 3, pp. 610–623, 1984.

B. Yegnanarayana, J. Sreekanth, and A. Rangarajan, Waveform estimation using group delay processing, *IEEE Transactions on Acoustics, Speech and Signal Processing*, vol. 33, no. 4, pp. 832–836, 1985.

B. Yegnanarayana and H. A. Murthy, Significance of group delay functions in spectrum estimation, *IEEE Transactions on Signal Processing*, vol. 40, no. 9, pp. 2281–2289, 1992.

L. Ying, Phase Unwrapping, *Wiley Encyclopedia of Biomedical Engineering*, Wiley, 2006.

D. Zhu and K. K. Paliwal, *Product of Power Spectrum and Group Delay Function for Speech Recognition*, Proceedings of IEEE International Conference on Acoustics, Speech, and Signal Processing (ICASSP), vol. 1, pp. 125–128, 2004.

3

Phase Estimation Fundamentals

Josef Kulmer and Pejman Mowlaee

Graz University of Technology, Graz, Austria

3.1   Chapter Organization

In this chapter, we will present the fundamentals of phase estimation and its potential with regard to phase-aware speech processing. We start with key examples demonstrating the influence of the applied window function on estimating the phase of one sinusoid. The influence of window functions on phase estimation performance is examined. We extend this example to multiple harmonics in noise. The detailed analysis reveals that the phase estimation error variance is affected by the choice of the analysis window and the frequency distance, as well as the interaction of amplitude and phase between the underlying harmonics. We continue with a detailed presentation of the existing methods for phase estimation available in the literature. Through experiments we demonstrate the usefulness of the phase estimation methods for speech enhancement applications. The chapter concludes with a comparative study, evaluating the effectiveness of the state-of-the-art phase estimation methods in speech enhancement.

3.2   Phase Estimation Fundamentals

3.2.1   Background and Fundamentals

The problem of interest in many signal processing applications, including radar, spectrum estimation, and signal enhancement, is to detect a signal of interest in a noisy observation. The signal of interest is often represented as a sum of sinusoids characterized by their amplitude, frequency, and phase parameters. Since these parameter triplets suffice to describe the signal, the problem degenerates to the detection and estimation of the sinusoidal parameters. This topic has been widely addressed in the literature of signal detection (Van Trees 1968) and estimation (Kay 1993). While many previous studies have been focused on deriving estimators for amplitude and frequency of sinusoids in noise (for an overview, see, e.g., Stoica and Moses 2005), the issue of phase estimation has been less addressed. Reliable phase estimation for practical applications has not been adequately addressed, in particular for signal enhancement.

3.2.2 Key Examples: Phase Estimation Problem

3.2.2.1 Example 1: Discrete-Time Sinusoid

To reveal the phase structure of one sinusoid's frequency response we consider the real-valued sequence

$$x(n) = \cos(\omega_0 n + \phi), \tag{3.1}$$

with ω_0 as frequency and ϕ as phase shift. Application of the discrete-time Fourier transform (DTFT), defined as

$$X(e^{j\omega}) = \text{DTFT}(x(n)) = \sum_{n=-\infty}^{\infty} x(n)e^{-j\omega n}, \tag{3.2}$$

yields the following frequency domain representation of the sequence $x(n)$:

$$X(e^{j\omega}) = \pi e^{j\phi}\delta(\omega - \omega_0) + \pi e^{-j\phi}\delta(\omega + \omega_0), \tag{3.3}$$

with $\delta(\omega)$ denoting the Dirac delta function. As the cosine function is symmetric ($\cos(\omega_0 n) = \cos(-\omega_0 n)$), only the phase shift ϕ determines the phase response of $X(e^{j\omega})$:

$$\angle X(e^{j\omega}) = \begin{cases} \phi, & \omega = \omega_0, \\ -\phi, & \omega = -\omega_0. \end{cases} \tag{3.4}$$

The left column of Figure 3.1 represents the sequence $x(n)$ along time n followed by its DTFT representation with real $\text{Re}\{X(e^{j\omega})\}$ and imaginary $\text{Im}\{X(e^{j\omega})\}$ parts as well as magnitude $|X(e^{j\omega})|$ and phase $\angle X(e^{j\omega})$ response. We set $\omega_0 = 0.1 \cdot 2\pi$ and $\phi = -\pi/8$.

The result in (3.4) is valid for an observation range of $n \in]-\infty, \infty[$. In practice, only a subset of ω_0 is available for analysis. This limitation can be represented by introducing an analysis window function, which is multiplied by the sequence $x(n)$. The modest analysis window is the rectangular, also known as boxcar or uniform window that has the value 1 within the range of N_w and 0 outside,

$$w(n) = \begin{cases} 1, & |n| \le \frac{N_w-1}{2}, \\ 0, & \text{else,} \end{cases} \tag{3.5}$$

for odd N_w. This window is symmetric ($w(n) = w(-n)$) and has the zero phase property, yielding a real-valued DTFT of the analysis window:

$$W(e^{j\omega}) = \sum_{n=-\infty}^{\infty} w(n)e^{-j\omega n} = \sum_{n=-(N_w-1)/2}^{(N_w-1)/2} 1 e^{-j\omega n} = \frac{\sin\left(\frac{N_w\omega}{2}\right)}{\sin\left(\frac{\omega}{2}\right)}, \tag{3.6}$$

also known as the *Dirichlet kernel*. The middle column of Figure 3.1 illustrates a symmetric rectangular window in time and frequency domain with the length of $N_w = 21$ and a DTFT length of $N = 31$. The real part is equal to the Dirichlet kernel while the imaginary part is equal to zero due to the symmetry of the window $w(n)$. The phase response represents the sign of $W(e^{j\omega})$, and is dependent on ω and equal to zero within the mainlobe width.

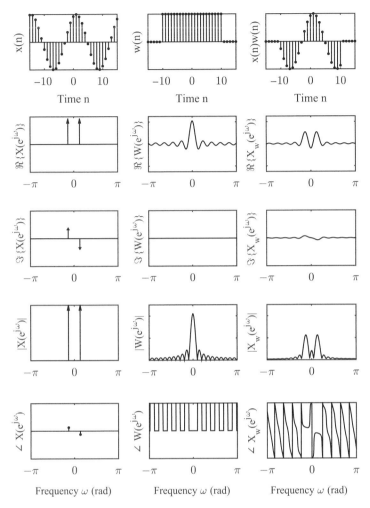

Figure 3.1 Visualization of the window impact on a sinusoid $x(n) = \cos(\omega_0 n + \phi)$ in time and frequency domains with $\omega_0 = 0.1 \cdot 2\pi$, $\phi = -\pi/8$, and a rectangular window with length $N_w = 21$. The window DTFT $W(e^{j\omega})$ is shifted dependent on ω_0 and multiplied by $e^{j\phi}$ and $e^{-j\phi}$, respectively, as shown in the phase response of DTFT $(x(n)w(n))$.

The product $x(n)w(n)$ corresponds to a convolution in the frequency domain according to

$$X_w(e^{j\omega}) = \mathrm{DTFT}(x(n)w(n)) = \frac{1}{2\pi}\{X * W\}(e^{j\omega}). \tag{3.7}$$

Plugging $X(e^{j\omega})$ and $W(e^{j\omega})$, derived in (3.3) and (3.6), respectively, into (3.7) yields the following expression:

$$X_w(e^{j\omega}) = \frac{1}{2}e^{j\phi}W(e^{j(\omega-\omega_0)}) + \frac{1}{2}e^{-j\phi}W(e^{j(\omega+\omega_0)}). \tag{3.8}$$

This is a rather important observation as the Dirichlet kernels are shifted along the frequency axis to $\omega = \omega_0$ and $\omega = -\omega_0$. The multiplication by the constants $e^{j\phi}$ and $e^{-j\phi}$, respectively, yields a complex-valued $X_w(e^{j\omega})$, as shown in the right column of Figure 3.1.

The terms on the right-hand side of (3.8) constructively add or eliminate each other, dependent on the values of ϕ and ω_0, described as the *leakage effect*.

The interaction between the Dirichlet kernels at $\omega = \omega_0$ is minimized if the frequency ω_0 fulfills the following requirement:

$$\omega_0 = \frac{2m\pi}{N_{\mathrm{w}}}, \quad m \in \mathbb{N}, \tag{3.9}$$

with m denoting the number of periods contained in one window length N_{w}. Figure 3.2 illustrates the impact of ω on the resulting magnitude and phase response of one

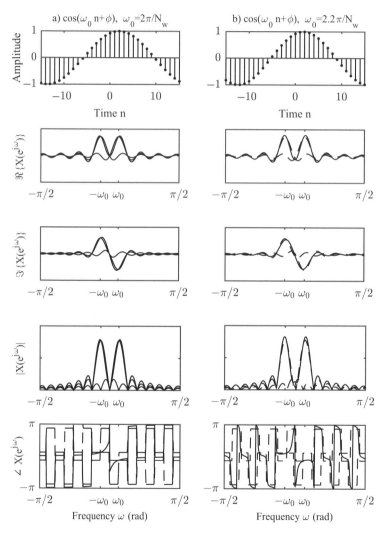

Figure 3.2 Relation of sinusoidal periods and window length and its impact on amplitude and phase. (a) A sinusoid multiplied by a boxcar window with a length of one period ($m = 1$). The Dirichlet kernels do not interfere at $\omega = \omega_0$ and $\omega = -\omega_0$, which yields an unbiased phase estimate of $\angle X_{\mathrm{w}}(e^{j\omega_0}) = \phi$. (b) The more general case of a window length that does not correspond to an integer multiplier of the sinusoid's period ($m = 1.1$). The amplitude as well as the phase do not approach the true value, and thus the outcome is biased.

sinusoid $x(n) = \cos(\omega_0 n + \phi)$ with $\phi = -\pi/8$, multiplied by a symmetric window of length $N_w = 31$. Setting $\omega_0 = 1\frac{2\pi}{N_w}$ leads to $m = 1$, hence for the phase response at frequency $\omega = \omega_0$ we obtain

$$X_w(e^{j\omega_0}) = \frac{1}{2}e^{j\phi}W(e^{j(\omega_0-\omega_0)}) + \frac{1}{2}e^{-j\phi}W(e^{j(\omega_0+\omega_0)}). \tag{3.10}$$

As the DTFT of the rectangular window $W(e^{j2\omega_0}) = 0$ for $m \in \mathbb{N}$, the phase response yields the true value of ϕ at frequency ω_0:

$$X_w(e^{j\omega_0}) = \frac{1}{2}e^{j\phi}CG, \tag{3.11}$$

$$\angle X_w(e^{j\omega_0}) = \phi, \tag{3.12}$$

defining $CG = W(e^{j0})$ as the coherent gain for the selected window[1]. The right column of Figure 3.2 demonstrates the more general case of $m \notin \mathbb{N}$. The sidelobes of $W(e^{j\omega})$ interact with each other, resulting in a biased phase response at ω_0. Note that the peak's location of the magnitude response is not at ω_0 due to the complex-valued superposition of both kernels. Therefore, any peak-picking method for obtaining a sinusoidal phase would result in a biased outcome.

In order to reduce the unpleasant impact of the sidelobe level, the choice of the window type becomes of particular interest. Basically, their behavior can be categorized by two characteristics: spectral leakage and frequency resolution. The frequency resolution is limited by the mainlobe width, which corresponds to the ability to resolve two adjacent spectral lines. To increase the frequency resolution a window function with a small mainlobe width is preferred. As a smaller mainlobe width is at the expense of a reduced sidelobe level the choice of an appropriate window function is a trade-off between a high frequency resolution and a low sidelobe level. Another way of optimizing the window choice is to adjust the window length N_w to fulfill the requirement in (3.9). However, adapting N_w needs knowledge of ω_0 and is in general only possible for one single sinusoid.

Figure 3.3 shows the influence of three prominent window types on the amplitude and phase response of the sequence $x_w(n) = \sin(\omega_0 n + \phi)w(n)$ with $\omega = 3.1 \cdot 2\pi/N_w$ and $\phi = -\pi/8$. So far, only the rectangular window was discussed. Its high frequency resolution (6 dB bandwidth of the mainlobe width: $\Delta\omega_{MW} = 1.21 \cdot 2\pi/N_w$) is at the cost of a rather poor sidelobe suppression level of -13 dB for the strongest neighboring sidelobe—Figure 3.3(a). The widely used Hamming window (b) consists of three shifted Dirichlet kernels with the purpose of minimizing the sidelobe levels, achieving a suppression of -42 dB at the cost of a worse frequency resolution (6 dB bandwidth of mainlobe: $\Delta\omega_{MW} = 1.81 \cdot 2\pi/N_w$). The amplitude and phase response of the Hamming-windowed sinusoid reveal the advantage of higher sidelobe suppression. The phase response at frequencies within the mainlobe width is determined by the true phase value $\phi = -\pi/8$. Compared to the rectangular window, the employment of a Hamming window results in a more robust phase estimation if an inaccurate frequency estimate of the sinusoid is given. Further, the magnitude's peak location is less shifted, which yields a more accurate phase estimation when using peak-picking. Finally, the Blackman window is presented in (c) with a sidelobe suppression of -58 dB and a 6 dB mainlobe bandwidth of $\Delta\omega_{MW} = 2.35 \cdot 2\pi/N_w$. The neighboring phase values at

---

1 See Harris (1978) for a detailed overview of the characteristics of and important definitions for window functions.

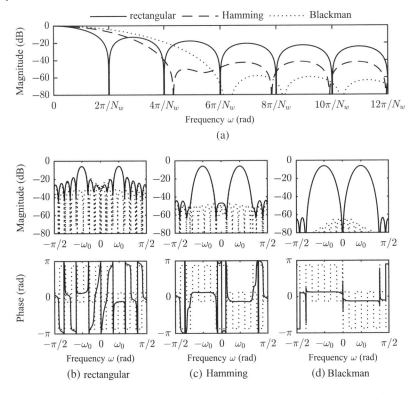

Figure 3.3 Illustration of three different windows' impacts on the magnitude and phase response of one sinusoid. The improved sidelobe suppression is at the cost of a higher mainlobe width resulting in a lower frequency resolution. For windows with higher sidelobe suppression, the phase response at frequency ω_0 is increasingly dominated by the phase ϕ within the mainlobe width.

ω_0 are strongly influenced by the true phase value of ϕ. However, both Hamming and Blackman windows deal with an increased mainlobe width. Once we extend our signal to multiple sinusoids, the mainlobe width plays a major role in selecting an appropriate window. If the mainlobe width contains more than one sinusoid then it is no longer possible to resolve the phase values of the sinusoids.

3.2.2.2 Example 2: Discrete-Time Sinusoid in Noise

So far, one sinusoid without additive noise was considered. For practical scenarios, the signal of interest is composed of multiple sinusoids corrupted with noise, as shown in Figure 3.4. The problem to solve is the estimation of the sinusoidal phase ϕ_h given its amplitude and frequency denoted as A_h and ω_h, respectively. Following the harmonic model of a speech signal we assume that the sinusoidal frequency ω_h is constrained to be a multiple harmonic of a fundamental frequency, i.e. $\omega_h = h\omega_0$ with $h \in [1, \ldots, H]$ denoting the harmonic index:

$$x(n) = \sum_{h=1}^{H} A_h \cos(\omega_h n + \phi_h) + d(n), \tag{3.13}$$

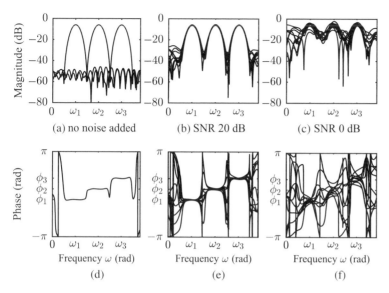

Figure 3.4 Illustration of the impact of additive white Gaussian noise on the magnitude and phase response of three neighboring sinusoids, windowed with Hamming without noise [(a),(d)], SNR = 20 dB [(b),(e)], and SNR = 0 dB [(c),(f)]. The left column shows the case if no noise is added. The mainlobes of the Hamming windows are sufficiently separated. The phase response shows that the neighboring frequencies of ω_1, ω_2, and ω_3 are dominated by the phase values of the sinusoids: $\phi_1 = -\pi/8$, $\phi_2 = \pi/8$, and $\phi_3 = 3\pi/8$. The middle and right columns present the impact of ten realizations of additive noise for SNR = 20 dB and SNR = 0 dB, respectively. With an increased noise level the phase values at the neighboring frequencies of ω_1, ω_2, and ω_3 are more affected by noise.

with $\omega_h = h2\pi f_0/f_s$ and $d(n)$ as the additive noise. Application of a window function $w(n)$, having non-zero values in the range $[-(N_w - 1)/2, (N_w - 1)/2]$, yields the windowed signal $x_w(n) = x(n)w(n)$. Similar to (3.8), the DTFT spectrum follows as

$$X_w(e^{j\omega}) = \sum_{h=1}^{H} \frac{A_h}{2} \left(e^{j\phi_h} W(e^{j(\omega - \omega_h)}) + e^{-j\phi_h} W(e^{j(\omega + \omega_h)}) \right) + D_w(e^{j\omega}), \tag{3.14}$$

where $D_w(e^{j\omega}) = \sum_{n=-(N_w-1)/2}^{(N_w-1)/2} d(n)w(n)e^{-j\omega n}$ is the windowed noise DTFT and $W(e^{j\omega})$ is the window frequency response. To have a better insight into the phase estimation problem, we are interested in the effect of the neighboring harmonics $h \neq \bar{h}$ on the desired harmonic $\bar{h}$. We evaluate the frequency response $X_w(e^{j\omega_{\bar{h}}})$ at the desired frequency $\omega_{\bar{h}}$,

$$X_w(e^{j\omega_{\bar{h}}}) = \frac{A_{\bar{h}}}{2} e^{j\phi_{\bar{h}}} \cdot CG + \frac{A_{\bar{h}}}{2} e^{-j\phi_{\bar{h}}} W(e^{j2\omega_{\bar{h}}})$$

$$+ \sum_{h=1, h \neq \bar{h}}^{H} \frac{A_h}{2} e^{j\phi_h} W(e^{j(\omega_{\bar{h}} - \omega_h)}) + \frac{A_h}{2} e^{-j\phi_h} W(e^{j(\omega_{\bar{h}} + \omega_h)}) + D_w(e^{j\omega_{\bar{h}}}). \tag{3.15}$$

The additive terms on the right-hand side of (3.15) show the interaction of the adjacent harmonics to the desired phase $\phi_{\bar{h}}$ as well as the impact of the additive noise. In the following, we are interested in the influence of these terms on the desired phase by rewriting (3.15) according to

$$X_w(e^{j\omega_{\bar{h}}}) = X_r(e^{j\omega_{\bar{h}}})e_c, \tag{3.16}$$

where $X_r(e^{j\omega_{\bar{h}}}) = \frac{A_{\bar{h}}}{2}e^{j\phi_{\bar{h}}}$ and e_c capture the phase estimation errors, given by

$$
e_c = \sum_{h=1}^{H} \frac{A_h}{A_{\bar{h}}} \left(e^{j(\phi_h - \phi_{\bar{h}})} W(e^{j(\omega_{\bar{h}} - \omega_h)}) + e^{-j(\phi_h + \phi_{\bar{h}})} W(e^{j(\omega_{\bar{h}} + \omega_h)}) \right)
$$

$$
+ \frac{2}{A_{\bar{h}}} e^{-j\phi_{\bar{h}}} D_w(e^{j\omega_{\bar{h}}}). \tag{3.17}
$$

In order to get more insight on the phase error term e_c, in the following we derive its phase mean and variance.

3.2.2.2.1 First Moment of e_c The mean value of the phase error term is given by

$$
\mathbb{E}_\phi(e_c) = \int_{-\pi}^{\pi} e_c\, p(\phi) d\phi, \tag{3.18}
$$

with $p(\phi)$ denoting the phase distribution. Applying (3.18) to (3.17) the first moment of e_c is given by

$$
\mathbb{E}_{\phi_{\bar{h}}}(e_c) = \sum_{h=1}^{H} \frac{A_h}{A_{\bar{h}}} \mathbb{E}_{\phi_{\bar{h}}}(e^{j(\phi_h - \phi_{\bar{h}})}) W(e^{j(\omega_{\bar{h}} - \omega_h)})
$$

$$
+ \sum_{h=1}^{H} \frac{A_h}{A_{\bar{h}}} \mathbb{E}_{\phi_{\bar{h}}}(e^{-j(\phi_h + \phi_{\bar{h}})}) W(e^{j(\omega_{\bar{h}} + \omega_h)}) \tag{3.19}
$$

$$
+ \frac{2}{A_{\bar{h}}} \mathbb{E}_{\phi_{\bar{h}}}(e^{-j\phi_{\bar{h}}}) D_w(e^{j\omega_{\bar{h}}}).
$$

3.2.2.2.2 Second Moment of e_c The second moment of the error term is given by

$$
\mathbb{E}_{\phi_{\bar{h}}}(e_c e_c^*) = \sum_{h_1=1}^{H-1} \sum_{h_2=h_1+1}^{H} C_1(\bar{h}, h_1, h_2) \mathbb{E}_{\phi_{\bar{h}}} (\cos(\phi_{h_1} - \phi_{h_2} + \angle w_1(\bar{h}, h_1, h_2)))
$$

$$
+ \sum_{h_1=1}^{H-1} \sum_{h_2=h_1+1}^{H} C_2(\bar{h}, h_1, h_2) \mathbb{E}_{\phi_{\bar{h}}} (\cos(\phi_{h_1} - \phi_{h_2} - \angle w_2(\bar{h}, h_1, h_2)))
$$

$$
+ \sum_{h_1=1}^{H} \sum_{h_2=1}^{H} C_3(\bar{h}, h_1, h_2) \mathbb{E}_{\phi_{\bar{h}}} (\cos(\phi_{h_1} + \phi_{h_2} + \angle w_3(\bar{h}, h_1, h_2)))
$$

$$
+ \sum_{h=1}^{H} C_4(\bar{h}, h) \mathbb{E}_{\phi_{\bar{h}}} \left(\cos(\phi_h - \phi_{\bar{h}} + \angle W(e^{j(\omega_{\bar{h}} - \omega_h)}) D_w(e^{-j\omega_{\bar{h}}})) \right)
$$

$$
+ \sum_{h=1}^{H} C_5(\bar{h}, h) \mathbb{E}_{\phi_{\bar{h}}} \left(\cos(\phi_h + \phi_{\bar{h}} - \angle W(e^{j(\omega_{\bar{h}} + \omega_h)}) D_w(e^{-j\omega_{\bar{h}}})) \right)
$$

$$
+ \frac{4}{A_{\bar{h}}^2} |D_w(e^{j\omega_{\bar{h}}})|^2 + C_6(\bar{h}), \tag{3.20}
$$

with the abbreviations

$$w_1(\overline{h}, h_1, h_2) = W(e^{j(\omega_{\overline{h}} - \omega_{h_1})}) W(e^{-j(\omega_{\overline{h}} - \omega_{h_2})}),$$

$$w_2(\overline{h}, h_1, h_2) = W(e^{j(\omega_{\overline{h}} + \omega_{h_1})}) W(e^{-j(\omega_{\overline{h}} + \omega_{h_2})}), \tag{3.21}$$

$$w_3(\overline{h}, h_1, h_2) = W(e^{j(\omega_{\overline{h}} - \omega_{h_1})}) W(e^{-j(\omega_{\overline{h}} + \omega_{h_2})}),$$

and the phase-independent constants

$$C_1(\overline{h}, h_1, h_2) = 2 \frac{A_{h_1} A_{h_2}}{A_{\overline{h}}^2} |w_1(\overline{h}, h_1, h_2)|,$$

$$C_2(\overline{h}, h_1, h_2) = 2 \frac{A_{h_1} A_{h_2}}{A_{\overline{h}}^2} |w_2(\overline{h}, h_1, h_2)|,$$

$$C_3(\overline{h}, h_1, h_2) = 2 \frac{A_{h_1} A_{h_2}}{A_{\overline{h}}^2} |w_3(\overline{h}, h_1, h_2)|,$$

$$C_4(\overline{h}, h) = 4 \frac{A_h}{A_{\overline{h}}^2} |W(e^{j(\omega_{\overline{h}} - \omega_h)})| \|D_w(e^{j\omega_{\overline{h}}})\|,$$

$$C_5(\overline{h}, h) = 4 \frac{A_h}{A_{\overline{h}}^2} |W(e^{j(\omega_{\overline{h}} + \omega_h)})| \|D_w(e^{j\omega_{\overline{h}}})\|,$$

$$C_6(\overline{h}) = \sum_{h=1}^{H} \frac{A_h^2}{A_{\overline{h}}^2} \left(|W(e^{j(\omega_{\overline{h}} + \omega_h)})|^2 + |W(e^{j(\omega_{\overline{h}} - \omega_h)})|^2 \right). \tag{3.22}$$

The second moment of the phase error in (3.20) provides useful insights on how the phase of the desired frequency $\omega_{\overline{h}}$ is a function of the chosen window, the additive noise, and the neighboring harmonics. Subsequently, the key factors are summarized as:

- $W(e^{j\omega})$: the magnitude and phase response of the analysis window function.
- $D_w(e^{j\omega})$: the magnitude and phase response of the windowed additive noise.
- $\frac{A_h}{A_{\overline{h}}}$: the amplitude ratio of the adjacent and desired harmonics.

The impact of the selected window $W(e^{j\omega})$ can be considered for two cases: First, where the harmonics are separated sufficiently, which means there is no neighboring harmonic within the mainlobe width. The window's amplitude $W(e^{j(\omega_{\overline{h}} - \omega_h)})$ for $\overline{h} \neq h$ suppresses the neighboring harmonic by its sidelobe level (see Figure 3.3), which results in a low impact of the neighboring harmonics on $\hat{\phi}_{\overline{h}}$. If the harmonics are not separated, i.e., the adjacent harmonic is located within the mainlobe width, then the phase error gets larger. The adjacent harmonic is not attenuated by the sidelobe level of the window and thus the phase estimation gets more biased. Figure 3.5 illustrates the impact of the adjacent harmonics for a Hamming and a Blackman window. The harmonics are sufficiently separated as the adjacent harmonics are outside of the mainlobe width. The Hamming window has a poorer sidelobe level which potentially yields a higher phase error. The larger sidelobe level of the Blackman window reduces the phase error but it also demands a longer window length N_w in order to separate the harmonics. The drawback of a longer window length is a reduced time resolution in the STFT spectral analysis. The phase estimation problem is therefore adequately addressed by choosing

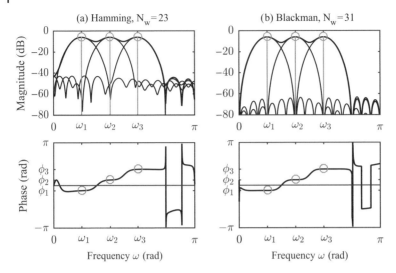

Figure 3.5 Spectral magnitude for a signal composed of three harmonics using (a) Hamming and (b) Blackman windows. The harmonics $h = 1, 2, 3$ are located at ω_1, ω_2, and ω_3. The broader mainlobe width of the Blackman window demands a higher window length of $N_w = 31$ in order to suppress the neighboring harmonics.

the best window that balances a good trade-off between a low phase variance and good enough time resolution.

3.2.3 Phase Estimation

The previous section showed the impact of neighboring harmonics and additive noise on the desired phase. These impacts play an important role in phase estimation and should be carefully considered in choosing a proper phase estimation method. In the following, two methods of phase estimation are presented: *maximum likelihood* (ML) and maximum *a posteriori* (MAP). Both are derived for a sum of harmonics plus additive white Gaussian noise.

3.2.3.1 Maximum Likelihood Estimation

The ML estimate of the sinusoidal phase in noise was proposed in Kay (1993). The ML estimator of phase is biased, hence contributes to poor performance at low signal-to-noise ratios. In particular, Peters amd Kay (2004) proved that no unbiased estimator of the phase of a sinusoid exists. Instead, they proposed estimators with less bias compared to the maximum likelihood estimator.

The phase estimation problem is formulated as estimation of the desired phase $\phi_{\overline{h}}$ for harmonic $\overline{h}$ among other harmonics, all embedded in noise:

$$x(n) = \sum_{h=1}^{H} A_h \cos(h\omega_0 n + \phi_h) + d(n), \tag{3.23}$$

where $d(n)$ is zero mean, white Gaussian noise with variance σ_d^2, A_h refers to the hth sinusoidal amplitude, and ω_0 is the fundamental frequency assumed to be known.

Stacking the independent observations $x(n)$ and phase values ϕ_h in the vectors $\mathbf{x} = [x(0), x(1), \ldots, x(N_w - 1)]$ and $\boldsymbol{\phi} = [\phi_1, \phi_2, \ldots, \phi_H]$, the likelihood function of the phase is given by:

$$p(\mathbf{x}; \boldsymbol{\phi}) = \prod_{n=0}^{N_w-1} p(x(n); \boldsymbol{\phi})$$

$$= \frac{1}{(2\pi\sigma_d^2)^{\frac{N}{2}}} \exp\left\{ \frac{-1}{2\sigma_d^2} \sum_{n=0}^{N_w-1} \left(x(n) - \sum_{h=1}^{H} A_h \cos(h\omega_0 n + \phi_h) \right)^2 \right\}. \qquad (3.24)$$

Applying the logarithm and neglecting the constants yields

$$L(\boldsymbol{\phi}) = \frac{1}{2\sigma_d^2} \sum_{n=0}^{N_w-1} \left(x(n) - \sum_{h=1}^{H} A_h \cos(h\omega_0 n + \phi_h) \right)^2, \qquad (3.25)$$

and the derivative with respect to $\phi_{\bar{h}}$ is given by

$$\frac{dL(\boldsymbol{\phi})}{d\phi_{\bar{h}}} = \frac{A_{\bar{h}}}{\sigma_d^2} \sum_{n=0}^{N_w-1} \left(x(n) - \sum_{h=1}^{H} A_h \cos(h\omega_0 n + \phi_h) \right) \sin(\bar{h}\omega_0 n + \phi_{\bar{h}}). \qquad (3.26)$$

Solving for the desired harmonic $\bar{h}$ we get

$$\sum_{n=0}^{N_w-1} x(n) \sin(\bar{h}\omega_0 n + \phi_{\bar{h}}) = \sum_{n=0}^{N_w-1} \left(\sum_{h=1}^{H} A_h \cos(h\omega_0 n + \phi_h) \right) \sin(\bar{h}\omega_0 n + \phi_{\bar{h}}). \qquad (3.27)$$

The term on the right-hand side of (3.27) is independent of the observation $x(n)$. This is of particular interest as it reveals the impact of the neighboring sinusoids $h \neq \bar{h}$ on the desired phase estimate $\phi_{\bar{h}}$. Re-writing the right-hand side of (3.27) gives more insight:

$$\sum_{n=0}^{N_w-1} \left(\sum_{h=1}^{H} A_h \cos(h\omega_0 n + \phi_h) \right) \sin(\bar{h}\omega_0 n + \phi_{\bar{h}})$$

$$= \frac{1}{2} \sum_{n=0}^{N_w-1} A_{\bar{h}} \sin(2\bar{h}\omega_0 n + 2\phi_{\bar{h}}) \qquad (3.28)$$

$$+ \frac{1}{2} \sum_{n=0}^{N_w-1} \sum_{h=1, h\neq\bar{h}}^{H} A_h \sin((\bar{h} + h)\omega_0 n + \phi_h + \phi_{\bar{h}})$$

$$+ \frac{1}{2} \sum_{n=0}^{N_w-1} \sum_{h=1, h\neq\bar{h}}^{H} A_h \sin((\bar{h} - h)\omega_0 n - \phi_h + \phi_{\bar{h}}),$$

with $\sin(-x) = -\sin(x)$. According to Kay (1993), the first term on the right-hand side of (3.28) can be approximated to zero if the desired harmonic frequency $\bar{h}\omega_0$ is not close to

0 or π. This can be stated by reformulating the term as a geometric series and evaluating the partial sum, leading to the following expression:

$$\frac{1}{2}\sum_{n=0}^{N_w-1} A_{\bar{h}} \sin(2\bar{h}\omega_0 n + 2\phi_{\bar{h}}) = \frac{A_{\bar{h}}}{2}\sum_{n=0}^{N_w-1} \mathrm{Im}\left\{e^{j(2\bar{h}\omega_0 n + 2\phi_{\bar{h}})}\right\}$$

$$= \frac{A_{\bar{h}}}{2}\mathrm{Im}\left\{e^{j2\phi_{\bar{h}}}\sum_{n=0}^{N_w-1}(e^{j2\bar{h}\omega_0})^n\right\}$$

$$= \frac{A_{\bar{h}}}{2}\mathrm{Im}\left\{e^{j2\phi_{\bar{h}}}\frac{1-e^{j2\bar{h}\omega_0 N_w}}{1-e^{j2\bar{h}\omega_0}}\right\}$$

$$= \frac{A_{\bar{h}}}{2}\mathrm{Im}\{e^{j2\phi_{\bar{h}}}e^{j\bar{h}\omega_0(N_w-1)}\}\frac{\sin(\bar{h}\omega_0 N_w)}{\sin(\bar{h}\omega_0)}$$

$$= \frac{A_{\bar{h}}}{2}\sin(\bar{h}\omega_0(N_w-1)+2\phi_{\bar{h}})\frac{\sin(\bar{h}\omega_0 N_w)}{\sin(\bar{h}\omega_0)}. \qquad (3.29)$$

The magnitude scales linearly with the harmonic's amplitude $A_{\bar{h}}$. Further, the Dirichlet kernel depends on the frequency $\bar{h}\omega_0$ only, and its magnitude is small for $\bar{h}\omega_0$ when it is not close to 0 or π. The Dirichlet kernel has its zeros at $\bar{h}\omega_0 = \frac{2m\pi}{N_w}$ with $m \in \mathbb{N}$, which is the special case of only full periods of sinusoids within the observation window of length N_w. At these frequencies the interference term in (3.29) goes to zero, and hence the leakage effect is minimized [see also (3.9)]. The second and third terms in (3.28) demonstrate the interaction between the harmonics. Both terms can be expressed similarly to (3.29) as

$$\frac{1}{2}\sum_{n=0}^{N_w-1}\sum_{h=1,h\neq\bar{h}}^{H} A_h \sin((\bar{h}\pm h)\omega_0 n \pm \phi_h + \phi_{\bar{h}}) =$$

$$\sum_{h=1,h\neq\bar{h}}^{H}\frac{A_h}{2}\sin\left((\bar{h}\pm h)\omega_0\frac{(N_w-1)}{2}\pm\phi_h+\phi_{\bar{h}}\right)\frac{\sin((\bar{h}\pm h)\omega_0\frac{N_w}{2})}{\sin((\bar{h}\pm h)\omega_0\frac{1}{2})}, \qquad (3.30)$$

showing that close neighboring harmonics $((\bar{h}\pm h)\omega_0 \ll)$ influence the phase estimation of the desired harmonic $\bar{h}$. In the case of $\bar{h}\omega_0 = \frac{2m\pi}{N_w}$ with $m \in \mathbb{N}$, the kernels have a zero which is equivalent to no interaction by the neighboring harmonic.

Assuming that the aforementioned conditions are met, the terms on the right-hand side of (3.28) can be approximated to zero and (3.27) becomes

$$\sum_{n=0}^{N_w-1} x(n)\sin(\bar{h}\omega_0 n + \phi_{\bar{h}}) \approx 0, \qquad (3.31)$$

which yields (3.27) to reduce to

$$\sum_{n=0}^{N_w-1} x(n)\sin(\bar{h}\omega_0 n)\cos(\phi_{\bar{h}}) + \sum_{n=0}^{N_w-1} x(n)\cos(\bar{h}\omega_0 n)\sin(\phi_{\bar{h}}) \approx 0, \qquad (3.32)$$

and finally the maximum likelihood harmonic phase estimate is given by

$$\hat{\phi}_{\overline{h}} = -\tan^{-1}\left(\frac{\sum_{n=0}^{N_w-1} x(n)\sin(\overline{h}\omega_0 n)}{\sum_{n=0}^{N_w-1} x(n)\cos(\overline{h}\omega_0 n)}\right). \tag{3.33}$$

It is interesting to note that the derived maximum likelihood phase estimate $\hat{\phi}_{\overline{h}}$ is equal to the phase of the DTFT response $X(e^{j\omega})$, evaluated at $\omega = \overline{h}\omega_0$.

3.2.3.2 Maximum a Posteriori Estimation

The ML phase estimator of the previous section is derived under the assumption that no prior knowledge of the phase distribution is available. In many cases, additional information is available, for example, in block processing prior information of the current time step can be gained from the past. In the following, a *maximum a posteriori* (MAP) phase estimator is derived that takes a prior distribution of phase into account. The employed *von Mises distribution* is the maximum entropy distribution for a given circular mean μ_h and concentration κ_h (Mardia and Jupp 2008, Section 3.5.4) and has been widely used in speech analysis/synthesis (Degottex and Erro 2014); it is given by

$$\phi_h \sim \mathcal{VM}(\mu_h, \kappa_h) \quad ; \quad p(\phi_h) = \frac{\exp\left(\kappa_h \cos(\phi_h - \mu_h)\right)}{2\pi I_0(\kappa_h)}, \tag{3.34}$$

where $I_\nu(\cdot)$ is the modified Bessel function of the first kind of order ν. The problem is formulated similar to (3.23): the noisy observation $x(n)$ consists of multiple harmonics in Gaussian noise $d(n)$ given by

$$x(n) = \sum_{h=1}^{H} A_h \cos(h\omega_0 n + \phi_h) + d(n), \tag{3.35}$$

with the likelihood function

$$p(\mathbf{x}|\boldsymbol{\phi}) = \frac{1}{(2\pi\sigma_d^2)^{\frac{N}{2}}} \exp\left\{\frac{-1}{2\sigma_d^2}\sum_{n=0}^{N_w-1}\left(x(n) - \sum_{h=1}^{H} A_h \cos(h\omega_0 n + \phi_h)\right)^2\right\}, \tag{3.36}$$

defining $\mathbf{x} = [x(0), x(1), \dots, x(N_w-1)]$ and $\boldsymbol{\phi} = [\phi_1, \phi_2, \dots, \phi_H]$. The MAP solution is obtained by solving

$$\hat{\boldsymbol{\phi}} = \arg\max_{\boldsymbol{\phi}} \frac{p(\mathbf{x}|\boldsymbol{\phi})p(\boldsymbol{\phi})}{p(\mathbf{x})} = \arg\max_{\boldsymbol{\phi}} p(\mathbf{x}|\boldsymbol{\phi})p(\boldsymbol{\phi}). \tag{3.37}$$

Plugging (3.34) and (3.36) into (3.37) and discarding the constants, we get:

$$L(\boldsymbol{\phi}) = \frac{-1}{2\sigma_d^2}\sum_{n=0}^{N_w-1}\left(x(n) - \sum_{h=1}^{H} A_h \cos(h\omega_0 n + \phi_h)\right)^2 + \sum_{h=1}^{H} \kappa_h \cos(\phi_h - \mu_h). \tag{3.38}$$

The MAP solution is given by taking the derivative of $L(\boldsymbol{\phi})$ with respect to the desired phase $\phi_{\overline{h}}$ and setting it equal to zero. Then, similar to (3.31), the equation can be approximated as

$$\frac{1}{\sigma_d^2}\sum_{n=0}^{N_w-1} A_{\overline{h}} x(n)\sin(\overline{h}\omega_0 n + \phi_{\overline{h}}) + \kappa_{\overline{h}}\sin(\phi_{\overline{h}} - \mu_{\overline{h}}) \approx 0 \tag{3.39}$$

for the desired phase $\phi_{\bar{h}}$. Using $\sin(a \pm b) = \sin(a)\cos(b) \pm \cos(a)\sin(b)$, we get

$$
\cos(\phi_{\bar{h}}) \left(\frac{A_{\bar{h}}}{\sigma_d^2} \sum_{n=0}^{N_w-1} (x(n)\sin(\bar{h}\omega_0 n)) - \kappa_{\bar{h}} \sin(\mu_{\bar{h}}) \right)
$$
$$
+ \sin(\phi_{\bar{h}}) \left(\frac{A_{\bar{h}}}{\sigma_d^2} \sum_{n=0}^{N_w-1} (x(n)\cos(\bar{h}\omega_0 n)) + \kappa_{\bar{h}} \cos(\mu_{\bar{h}}) \right) \approx 0,
$$

(3.40)

and the MAP solution is given by

$$
\hat{\phi}_{\bar{h}} = \tan^{-1} \left(\frac{-\frac{A_{\bar{h}}}{\sigma_d^2} \sum_{n=0}^{N_w-1} x(n)\sin(\bar{h}\omega_0 n) + \kappa_{\bar{h}} \sin(\mu_{\bar{h}})}{\frac{A_{\bar{h}}}{\sigma_d^2} \sum_{n=0}^{N_w-1} x(n)\cos(\bar{h}\omega_0 n) + \kappa_{\bar{h}} \cos(\mu_{\bar{h}})} \right).
$$

(3.41)

The MAP phase estimate is a function of the following parameters: the concentration $\kappa_{\bar{h}}$ and the circular mean $\mu_{\bar{h}}$ of the von Mises prior, the observation length N_w, and a term proportional to the local signal-to-noise ratio $\frac{A_{\bar{h}}}{\sigma_d^2}$. As an extreme scenario of large SNRs ($A \gg \sigma_d^2$), the MAP estimator asymptotically degenerates to the ML estimate (3.33) given by the noisy DTFT phase sampled at the harmonic frequency. In a high SNR scenario the noisy phase is more weighted than the von Mises prior, whereas at lower SNRs the phase estimate is dominated by $\mu_{\bar{h}}$. A low $\kappa_{\bar{h}}$ yields a uniform phase prior while a large concentration $\kappa_{\bar{h}} \to \infty$ resembles a Dirac delta denoting a high certainty in the prior phase mean value. It can be concluded that the von Mises distribution phase prior allows for a flexible framework that takes into account the uncertainty in phase estimation. A performance evaluation and comparison of the derived ML and MAP estimators are presented in Section 3.4.1.

3.3 Existing Solutions

In this section we provide a review of the existing phase estimators that have been proposed in the literature. Table 3.1 lists these phase estimators with their references. In the following, the details for each method and the central underlying ideas used will be elaborated.

3.3.1 Iterative Signal Reconstruction

3.3.1.1 Background

The problem of estimating the phase spectrum from an observed signal dates back to the 1980s, where researchers made the first steps towards recovering a time domain signal from just its STFT magnitude. For example, Hayes *et al.* (1980) proposed techniques to reconstruct a signal from its magnitude or phase information only. Quatieri and Oppenheim (1981) derived an iterative solution for reconstructing a time domain representation of minimum-phase signals. Alsteris and Paliwal (2007) presented an overview of iterative, one-dimensional signal reconstruction methods from the magnitude or phase short-time Fourier spectra. Sturmel and Daudet (2011) presented a review of techniques for signal reconstruction without phase information, i.e., only from the spectrogram of the signal. As the magnitude spectrum alone cannot determine the zeros, and therefore

Table 3.1 Categorization of phase estimation methods with citations.

Phase estimator	Reference
Iterative signal reconstruction	Griffin and Lim (1984)
Randomization	Sugiyama and Miyahara (2013)
Geometry-based with constraints	Mowlaee *et al.* (2012); Mowlaee and Saeidi (2014)
Temporal phase continuity	Mehmetcik and Ciloglu (2012)
STFT phase improvement across frequency (STFTPI)	Krawczyk and Gerkmann (2014)
Least squares	Chacon and Mowlaee (2014)
Smoothing of unwrapped phase	Kulmer *et al.* (2014); Mowlaae and Kulmer (2015b)
Maximum *a posteriori* (MAP)	Kulmer and Mowlaee (2015a)

the phase spectrum cannot be determined by the magnitude spectrum, the uniqueness of a possible solution is of interest. In particular, following McGowan and Kuc (1982), the Fourier phase spectrum of a signal of length N_{w}, for a known magnitude spectrum scenario, allows for a unique decision amongst $2^{N_{\mathrm{w}}-1}$ possible time domain signals. In the following, we present some background on this concept and continue with the well-known Griffin–Lim iterative signal reconstruction method (Griffin and Lim 1984).

Nawab *et al.* (1983) listed the following required conditions under which a unique signal reconstruction from a given magnitude STFT is possible:

- The window function $w(n)$ is known and non-zero for $n \in [0, N_{\mathrm{w}} - 1]$.
- The overlap of consecutive frames is at least 50%.
- The signal is one-sided.
- The number of successive zero samples in the signal should be less than the frame shift S.

As an alternative approach for signal reconstruction, simultaneous extrapolation was proposed by Griffin and Lim (1984). In the next section, we will continue with the Griffin–Lim method of reconstructing a signal from partial information due to its high popularity and practical use.

3.3.1.2 Griffin–Lim Algorithm (GLA)

Griffin and Lim (1984) derived an iterative algorithm to estimate a signal from its modified short-time Fourier transform. Let $x_w(lS, n)$ denote the windowed segment of the signal $x(n)$, where the window is shifted by l times the frameshift S, with $n \in [0, N_{\dot{w}} - 1]$. Let $Y_w(lS, e^{j\omega}) = \mathrm{DTFT}\{y_w(lS, n)\}$ be a given modified STFT (MSTFT), as the Fourier transform of $y_w(lS, n)$ with respect to n given by

$$y_w(lS, n) = \frac{1}{2\pi} \int_{\omega=-\pi}^{\pi} Y_w(lS, e^{j\omega}) e^{j\omega n} d\omega. \tag{3.42}$$

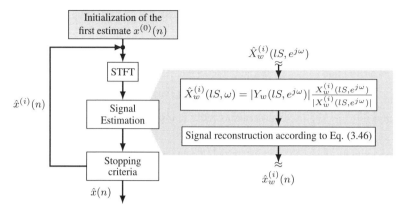

Figure 3.6 Iterative framework for signal reconstruction showing the GLA update procedure to reconstruct the signal at iteration index *i*, following Griffin and Lim, (1984).

The goal of the Griffin–Lim algorithm is to find an estimated sequence $x(n)$ whose STFT $X_w(lS, e^{j\omega})$ is closest to $Y_w(lS, e^{j\omega})$ in the sense of the squared error. Considering the squared error distance measure between the two spectra and using Parseval's theorem, we get:

$$D_m(x(n), Y_w(lS, e^{j\omega})) = \sum_{l=-\infty}^{\infty} \sum_{n=-\infty}^{\infty} [x_w(lS, n) - y_w(lS, n)]^2. \tag{3.43}$$

Setting the gradient with respect to $x(n)$ to zero and solving for $x(n)$ results in:

$$x(n) = \frac{\sum_{l=-\infty}^{\infty} w(lS - n) y_w(lS, n)}{\sum_{l=-\infty}^{\infty} w^2(lS - n)}. \tag{3.44}$$

This solution is referred to as *least squares error estimation* (LSEE) from the MSTFT, which is similar to the standard overlap-add procedure (Rabiner and Schafer 1978; Allen 1977) given by:

$$x(n) = \frac{\sum_{l=-\infty}^{\infty} y_w(lS, n)}{\sum_{l=-\infty}^{\infty} w(lS - n)}. \tag{3.45}$$

The overlap outcome in (3.45) is different from the LSEE-MSTFT solution in (3.44) in two ways: (i) the sequence $y_w(n)$ is windowed before applying the overlap-add, and (ii) the normalization factor $w(lS - n)$ is squared. It is also important to note that the solution in (3.44) was derived for the purpose of estimating a signal from its MSTFT in a least squares sense. Equation (3.45), in contrast, was proposed to reconstruct a signal from its exact STFT.

As the gradient of D_m is non-positive, the iterative technique reduces the distance D_m for every iteration *i*. To extend the least squares solution in (3.44) to an iterative framework, we define $\hat{x}^{(i)}(n)$ as the estimated $x(n)$ at the *i*th iteration. The estimated signal at the $(i + 1)$th iteration denoted by $\hat{x}^{(i+1)}(n)$ is then given by taking the STFT of $\hat{x}^{(i)}(n)$, replacing its magnitude spectrum with $Y_w(lS, e^{j\omega})$, and repeatedly applying (3.44) to find the time domain signal with the STFT most similar to this modified STFT, as illustrated in Figure 3.6. This results in the following update equation:

$$\hat{x}^{(i+1)}(n) = \frac{\sum_{l=-\infty}^{\infty} w(lS - n) \frac{1}{2\pi} \int_{\omega=-\pi}^{\pi} \hat{X}_w^{(i)}(lS, e^{j\omega}) e^{j\omega n} d\omega}{\sum_{l=-\infty}^{\infty} w^2(lS - n)}, \tag{3.46}$$

where we have

$$\hat{X}_w^{(i)}(lS, e^{j\omega}) = |Y_w(lS, e^{j\omega})| \frac{X_w^{(i)}(lS, e^{j\omega})}{|X_w^{(i)}(lS, e^{j\omega})|}. \tag{3.47}$$

Griffin and Lim set $\hat{X}_w^{(i)}(lS, e^{j\omega}) = |Y_w(lS, e^{j\omega})|$ for the case of $X_w^{(i)}(lS, e^{j\omega}) = 0$.

3.3.1.3 Extensions of the GLA

The Griffin–Lim algorithm has the following drawbacks:

1) Its convergence is slow in terms of the number of iterations required and the computation time per iteration.
2) It does not perform any local optimization in order to improve the MSTFT's consistency.
3) Its computation requires offline processing as it involves STFT and inverse STFT procedures at every iteration.
4) GLA does not provide a good initial estimate since it neglects the available information given by cross-frame dependencies. This may lead to inconsistency in the resulting STFT.

The quality of the reconstructed signal using the GLA is highly influenced by the parameters used in the STFT frame setup (window length, window type, and frame shift) as well as the characteristics of the input signal. Artifacts in the form of reverberated sound and pre-echo occur. Inspired by the Griffin–Lim iterative signal reconstruction, several different variants have been proposed. In the following, we review some techniques that have been proposed to improve GLA. Other extensions of the GLA for the application of single-channel source separation will be discussed in detail in Chapter 5.

3.3.1.3.1 Real-time Iterative Magnitude Spectrogram Inversion (RTISI) GLA requires the magnitude spectra of all frames of the input data. Therefore, it is not appropriate for real-time applications. To resolve this, Zhu *et al.* proposed two practical implementations: real-time iterative magnitude spectrogram inversion (RTISI) (Zhu *et al.* 2007) and the RTISI with look-ahead (RTISI-LA) algorithm (Zhu *et al.* 2006). RTISI aims to improve the original GLA by allowing for an online implementation as well as generating better initial phase estimates. At each frame the algorithm uses information about the current and the previous frames within the overlap region. The phase of the signal reconstructed from the previous frame is used as an initial phase estimate for the current frame iteration. While RTISI does not include future frames, RTISI-LA implements look-ahead where GLA updates are carried out on the phases of multiple frames. Both RTISI and RTISI-LA reduce the computational load while maintaining signal reconstruction performance at a reasonable level. Both methods successfully overcome GLA's poor initialization and its requirement for offline processing.

3.3.1.3.2 Spectrogram Consistency Griffin and Lim reported improved performance for large overlaps, for example 87.5% for the specific choice of the square-root Hann window (Griffin and Lim 1984). This allows for exploiting correlation between the magnitude and phase spectra of the short-time Fourier transformed frames. In fact, the key idea behind signal reconstruction from partial STFT information (either magnitude or phase) is the redundancy in STFT analysis. Applying a minimum of 50% overlap, the magnitude and phase spectra are not independent since a real input signal of length N_w results in $2N_w$ real coefficients (neglecting the boundary effects of the selected window).

The redundancy of the STFT representation is expressed by a criterion called consistency criterion Le Roux *et al.* (2010a). Overlapping frames imply spectro-temporal correlations amongst them. By modifying the magnitude spectrum only, the correlations are mitigated. The STFT of the resynthesized MSTFT is again redundant. Therefore, it is possible to introduce a criterion (consistency) that describes the degree of correlatedness of an MSTFT. Let $\mathbf{X}$ denote the spectrogram matrix for all frames l and frequency bins k, then the inconsistency operator $\mathcal{I}(\cdot)$ is defined by

$$\mathcal{I}(\mathbf{X}) = \mathbf{X} - \text{STFT}(\text{iSTFT}(\mathbf{X})) = \mathbf{X} - \mathcal{G}(\mathbf{X}), \tag{3.48}$$

with $\mathcal{G}(\mathbf{X}) = \text{STFT}(\text{iSTFT}(\mathbf{X}))$. This implies that in the case of a consistent spectrogram ($\mathbf{X} = \mathcal{G}(\mathbf{X})$) the constraint $\mathcal{I}(\mathbf{X}) = 0$ is fulfilled. The inconsistency operator in (3.48) is used to enforce temporal coherence of a given signal. It is important to note that in many practical applications (including speech enhancement, time-scale modification, and signal modification), the spectrogram used for signal reconstruction might not belong to the set of consistent spectrograms. The goal is then to find the closest estimate that minimizes the norm of $\mathcal{I}(\mathbf{X})$. Figure 3.7 demonstrates the concept of spectrogram consistency.

Le Roux *et al.* (2010b) modified the GLA to speed up its convergence. They proposed to take the sparseness of the input signal into account by limiting the GLA update routine to those bins with a significant magnitude. The rationale behind this choice is the fact that the bins with low amplitudes contribute less in producing undesired artifacts in GLA. Finally, they proposed an *on-the-fly* update scheme where the newly updated values are used for the computation of the subsequent updates of the other bins. In contrast to the *stepwise update* used in GLA, which involves updates from the previous step

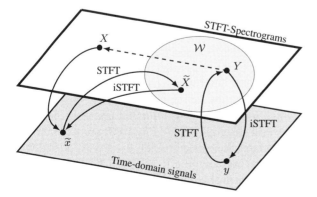

Figure 3.7 Spectrogram consistency concept used in Griffin–Lim iterative signal reconstruction. A consistent spectrogram (belonging to the set **W**) verifies **Y** = **G(Y)**, while for an inconsistent spectrogram **X** ≠ $\mathcal{G}$(**X**) leading to $\mathcal{I}$(**X**) ≠ 0.

only, a faster convergence is achieved using their proposed on-the-fly update scheme. This reduces the computation cost and requires less memory space.

3.3.1.3.3 FGLA Perraudin *et al.* (2013) presented a faster version of the GLA with lower computational complexity and higher accuracy. They formulated the phase recovery problem in an optimization framework as follows: From a given set of non-negative coefficients $|\mathbf{Y}|$ find a signal $\mathbf{x}$ such that the magnitude of its STFT is as close as possible to $|\mathbf{Y}|$. Similar to the GLA, they chose the l_2 norm as a measure of closeness. For real positive coefficients $|\mathbf{Y}|$, the solution $\mathbf{x}^*$ is found by

$$x^*(n) = \arg\min_{x(n)} \||\text{STFT}x(n)| - |\mathbf{Y}|\|_2, \tag{3.49}$$

where $|\mathbf{Y}|$ is a valid STFT magnitude only if there exists an $\mathbf{x}$ such that $|\text{STFT}(\mathbf{x})| = |\mathbf{Y}|$. In Experiment 3.3 in Section 3.4.3 we will demonstrate the performance of the FGLA in comparison to the GLA.

3.3.2 Phase Reconstruction Across Time

The following methods operate mainly in the DFT domain with discrete frequency bin k where for a shorter notation we change the arguments $e^{j\omega_k}$ to its index k. Mehmetcik and Ciloglu (2012) proposed to estimate the clean spectral phase using the phase continuity constraint across time. Assume that the phase of the current frame l, $\hat{\phi}_x(k, l)$, is desired and that we have access to the estimated phase from the previous frame $l - 1$, $\hat{\phi}_x(k, l - 1)$. Mehmetcik and Ciloglu followed the phase vocoder principle (Dolson 1986) and proposed the prediction of the phase across time for consecutive frames as illustrated in Figure 3.8. The predicted phase value at frame l and frequency bin k is given by

$$\hat{\phi}_x(k, l) = \hat{\phi}_x(k, l - 1) + \alpha_i \text{IF}(k, l), \tag{3.50}$$

where $\text{IF}(k, l)$ is the *instantaneous frequency* (IF) defined in Chapter 2 and α_i denotes a positive real-valued factor that indicates an increasing or decreasing IF (phase vocoder case for $\alpha_i = 1$). Figure 3.8 shows the block diagram for the phase reconstruction method across time (Mehmetcik and Ciloglu 2012). It is important to note that the accuracy of the temporal phase reconstruction is heavily dependent on the stationarity of the instantaneous frequency, i.e., how much the fundamental frequency changes within the

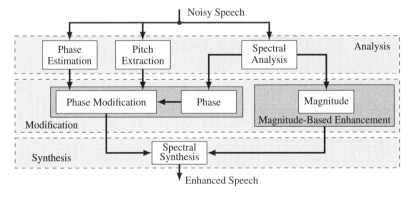

Figure 3.8 Speech enhancement by maintaining phase continuity and phase reconstruction across time as proposed in Mehmetcik and Ciloglu (2012) for voice frames.

duration of the two consequent frames. As a result, the performance of the method is strongly affected by the pitch estimation accuracy (Mehmetcik and Ciloglu 2012).

Phase estimation across time has also been proposed by Kleinschmidt *et al.* (2011) in the context of complex spectral subtraction for robust automatic speech recognition (they called it *phase estimation with delay projection*). It relies on the predictability of the phase at each time–frequency bin from past frames. In practice, the reference phase is selected from the phase of the most recent speech frame. Kleinschmidt *et al.* termed their method *estimation based on sinusoid stationarity* and showed improved automatic speech recognition performance for a known clean phase spectrum.

3.3.3 Phase Reconstruction Across Frequency

In addition to restoring the phase structure across time, Krawczyk and Gerkmann (2014) proposed to make use of the sinusoidal model of speech in order to reconstruct its spectral phase structure across frequency. The method was termed *STFT phase improvement* (STFTPI). Let $\omega_h = 2\pi f_h / f_s$ be the normalized angular frequency of the hth harmonic. The authors assume that the phase at the STFT bin k is dominated by the closest sinusoid's phase defined as ω_h^k given by

$$\omega_h^k = \arg \min_{\omega_h}(|\omega_k - \omega_h|). \tag{3.51}$$

While $\omega_k N_{\text{DFT}}/2\pi = k$ is an integer, $\omega_h^k N_{\text{DFT}}/2\pi$ is not necessarily resolved by the DFT. Let $\phi_{\text{B}}(k, l)$ be the baseband STFT phase given by modulating each STFT band into the baseband, and we obtain the baseband STFT phase as

$$\phi_{\text{B}}(k, l) = \angle X(k, l)e^{-j\omega_k lS}, \tag{3.52}$$

with S denoting the frame shift. Given the phase information at frequency bin k', the phase for the frequency bins $k' + \Delta k'$ dominated by the spectral harmonic at frequency band k is given by:

$$
\begin{aligned}
\hat{\phi}_{\text{B}}(k' + \Delta k', l) = {} & \phi_{\text{B}}(k', l) - \Delta k' \frac{2\pi}{N_{\text{DFT}}} lS \\
& + \phi_w\left(k' + \Delta k' - \frac{\omega_h^k N_{\text{DFT}}}{2\pi}\right) - \phi_w\left(k' - \frac{\omega_h^k N_{\text{DFT}}}{2\pi}\right),
\end{aligned}
\tag{3.53}
$$

defining $\phi_w(\cdot)$ as the phase spectrum of the analysis window $w(n)$. STFTPI relies on a phase estimate at the signal segment l_0, denoting the onset of a voiced phoneme. Combining the reconstructed phase with the noisy spectral amplitude and demodulating it back to the kth frequency bin, the clean phase-enhanced speech estimate is finally given by:

$$\hat{X}(k, l) = (|Y_{\text{B}}(k, l)|e^{j\hat{\phi}_{\text{B}}(k,l)})e^{j\omega_k lS}, \tag{3.54}$$

where $Y_{\text{B}}(k, l)$ is the baseband representation of the noisy speech. The phase is only reconstructed at voiced segments while the noisy phase is assigned at unvoiced frames. The method imposes a strictly harmonic structure up to the Nyquist frequency. Similar to a comb filter, the method harmonizes the noisy speech via modification of the noisy phase spectrum across frequency. This procedure, depending on the noise type and the structure of the desired speech, may contribute to improved perceived speech quality, since in between harmonics the noise components are suppressed.

The method requires an accurate fundamental frequency estimate together with reliable voice activity detection. It is important to note that a wrong signal detection eventually results in buzzyness and reduced quality, compared to the level of the input noisy phase. A similar concept was employed in Patil and Gowdy (2014, 2015) to estimate noise between the harmonics. Patil and Gowdy proposed estimating the frame-level baseband phase difference (BPD)[2] of clean speech from a noisy speech. The method provided a more accurate noise estimation for highly non-stationary noise in the context of spectral subtraction speech enhancement. Improved PESQ scores were reported when this BPD-based noise estimate was fused with a speech enhancement method, e.g., MMSE-STSA (see Chapter 4), showing improved noise suppression and speech quality (Patil and Gowdy 2015).

3.3.4 Phase Randomization

Miyahara and Sugiyama proposed randomizing the phase as an alternative paradigm to mitigate the detrimental effect of noisy phase in speech enhancement (Sugiyama and Miyahara 2013; Miyahara and Sugiyama 2013, 2014). They considered the noise reduction in the context of auto-focusing noise, a mechanical noise in audio-visual recording using digital still cameras. In particular, when using the zoom function during a movie recording, the auto-focusing noise becomes stronger and possibly masks the desired speech (Miyahara and Sugiyama 2014). The main principle in phase randomization is to take into account the linear phase character of the underlying auto-focusing noise. Therefore, the structured patterns of the noise signal's phase are de-emphasized by randomizing the observed noisy phase spectrum. Phase randomization restricts the constructive addition (in-phase) of noise-dominated spectral components. The results in Miyahara and Sugiyama (2014) demonstrate that, by using the overlap-add at signal reconstruction, the consecutive segments with randomized phase brought improved noise reduction performance due to the removal of the linear noise phase pattern. The improvement was justified via subjective listening results using a comparative category rating (CCR) test (Sugiyama and Miyahara 2013; Miyahara and Sugiyama 2013, 2014).

The detailed design of the enhancement algorithm for auto-focus zoom noise suppression is shown in Figure 3.9, where phase randomization is highlighted. In particular, the

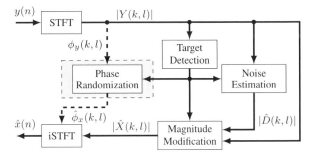

Figure 3.9 Block diagram for phase randomization proposed for zooming noise suppression (Sugiyama and Miyahara 2013).

---

2 See Chapter 2 for more details on BPD representation.

input noisy spectral coefficient $\phi_y(k, l)$ is randomized based on a randomization index $r(k, l)$ to obtain an enhanced signal phase $\hat{\phi}_x(k, l)$ given by (Miyahara and Sugiyama 2014):

$$\hat{\phi}_x(k, l) = \phi_y(k, l) + r(k, l)\phi_r(k, l), \tag{3.55}$$

where $\phi_r(k, l)$ is a random value within $[-\pi, \pi[$. For signal components detected as spectral peaks in the frequency domain the noisy spectral phase remains unaltered, hence $r(k, l) = 0$. For spectral components detected as non-target components, the noisy phase is replaced with a randomized value, hence $r(k, l) = 1$. Each spectral peak is considered to have a certain bandwidth and therefore is not an isolated frequency bin. All peaks are labeled with a peak flag if they lie above the estimated value of the auto-focusing noise level. These selected peaks are eventually used to decide where to apply phase randomization.

Correct knowledge about the local SNR is the key for the success of the phase randomization approach. To show this, in the following, we consider a proof-of-concept experiment. A clean speech utterance is corrupted with babble noise at 0 dB SNR. The corresponding spectrograms are shown in Figure 3.10. Phase randomization is applied in two ways: once with the correct local SNR defined as $\mathrm{SNR}(k, l) = \frac{|X(k,l)|^2}{|D(k,l)|^2}$, and once using its blind estimate at harmonics compared to an estimated noise PSD. A negligible improved PESQ score is obtained with phase randomization in the blind SNR scenario compared to the noisy speech. In contrast, with the correct SNR, the phase randomization results in significant improvement in perceived quality and speech intelligibility predicted by PESQ and short-time objective intelligibility (STOI), respectively. The difficulty of estimating a reliable local SNR limits the usefulness of the phase randomization approach in practice.

It is important to note that while the other phase estimators discussed so far aim to estimate the clean speech phase, the phase randomization scheme attempts to reduce the impact of the noise phase by randomizing its structure for signal reconstruction.

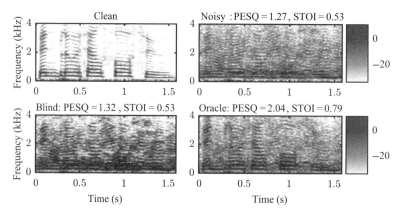

Figure 3.10 Proof-of-concept experiment for phase randomization: (top) clean versus noisy speech, (bottom) phase randomization with blind and oracle SNRs.

3.3.5 Geometry-Based Phase Estimation

Mowlaee *et al.* (2012) proposed to solve the phase estimation problem using a geometric approach. From geometry and the analytic solution for the signal's phase, two phase candidates are available at each time–frequency bin which differ in terms of the sign of the phase difference, denoted by $\Delta\phi(k,l) = \phi_x(k,l) - \phi_d(k,l)$ (see Figure 3.11). In the complex plane, the additivity for speech and noise spectra is fulfilled and we have

$$|Y(k,l)|e^{j\phi_y(k,l)} = |X(k,l)|e^{j\phi_x(k,l)} + |D(k,l)|e^{j\phi_d(k,l)}. \tag{3.56}$$

A valid solution for an estimated phase for either of the two signals should fulfill the constraint of a valid geometry. This constraint will be used as a minimum signal reconstruction error criterion in the derivation of the geometry-based method. In particular, from the geometry in the complex domain we rewrite (3.56) as

$$|Y(k,l)|e^{j\phi_y(k,l)} = (|X(k,l)| + |D(k,l)|e^{-j\Delta\phi(k,l)})e^{j\phi_x(k,l)}. \tag{3.57}$$

To fulfill this constraint, exact knowledge of $\Delta\phi(k,l)$ is required; this is ambiguous due to the parity in the sign of $\sin(\Delta\phi(k,l))$, and therefore a unique solution for (3.57) is not possible and some additional constraints on the spectral phase are required to select the correct phase value.

We define $\hat{\phi}_x^{(a)}(k,l)$ and $\hat{\phi}_d^{(a)}(k,l)$ as the sets of ambiguous phase estimates of speech and noise sources, respectively. Each phase set itself consists of two phase values: $\hat{\phi}_x^{(a)}(k,l) = \{\hat{\phi}_x^{(1)}(k,l), \hat{\phi}_x^{(2)}(k,l)\}$ and $\hat{\phi}_d^{(a)}(k,l) = \{\hat{\phi}_d^{(1)}(k,l), \hat{\phi}_d^{(2)}(k,l)\}$, for speech and noise signals, respectively. The pair of ambiguous phase values satisfy all observations regarding the noisy complex spectrum and the spectral amplitude of the underlying signals, while they only differ in their resulting sign in $\Delta\phi = \pm|\phi_x(k,l) - \phi_d(k,l)|$. The minimum reconstruction error criterion in (3.57) is applied to find the best pair of ambiguous phase values at the current time–frequency cell, defined as

$$|\hat{Y}(k,l)|e^{j\hat{\phi}_y(k,l)} = |X(k,l)|e^{j\hat{\phi}_x(k,l)} + |D(k,l)|e^{j\hat{\phi}_d(k,l)}, \tag{3.58}$$

$$e(k,l) = |Y(k,l) - |\hat{Y}(k,l)|e^{j\hat{\phi}_y(k,l)}|. \tag{3.59}$$

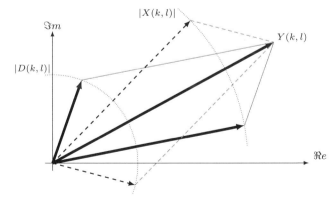

Figure 3.11 Representation of $Y(k,l)$ as the sum of speech $X(k,l)$ and noise $D(k,l)$ in the complex plane. Due to the parity in the sign of $\sin\Delta\phi$ in (3.57), there is an ambiguity in the set of phase candidates.

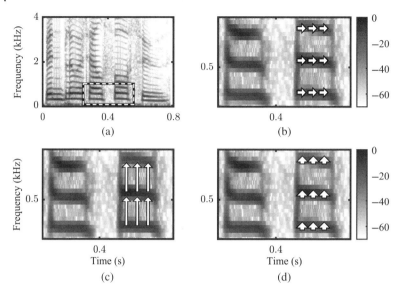

Figure 3.12 Phase constraints across time (IFD), harmonics (RPS), and frequency (GDD). The arrows show the coordination to which the proposed constraints are applied on the phase spectrum.

The geometry-based approach is applied to the spectral peaks with the hth spectral peak denoted by $\{k_h\}_{h=1}^{H}$ and H being the total number of peaks. Then the optimal phase values at each frame l are given by drawing all combinations from the ambiguous set $\hat{\phi}_x^{(a)}(k_h, l)$, given by

$$\hat{\phi}_x^*(k_h, l) = \underset{\hat{\phi}_x^{(a)}(k_h, l)}{\arg \min} \; d_a, \tag{3.60}$$

where d_a refers to an auxiliary constraint on the spectral phase, imposed to solve the phase ambiguity problem. Figure 3.12 graphically represents three constraints across time and frequency for a real speech signal. Mowlaee *et al.* applied a group delay deviation constraint (Mowlaee *et al.* 2012) and two other constraints: instantaneous frequency deviation (IFD) and relative phase shift (RPS)[3] (Mowlaee and Saeidi 2014) for the purpose of signal reconstruction for single-channel signal enhancement. By incorporating such additional constraints and evaluating the ambiguity set of phase candidates, the most likely phase estimate is used for signal reconstruction. The geometry-based phase estimators were successfully integrated into iterative phase-aware closed-loop single-channel speech enhancement.[4]

In the following, we will only present details of the group delay deviation constraints, while the derivation of the other constraints is similar in the framework of geometry-based phase estimation. From the group delay definition provided in Chapter 2, and assuming a short-time Fourier analysis, for a causal window of length symmetric around $\frac{N_w-1}{2}$ the group delay is a constant value $\tau_w = \frac{N_w-1}{2}$. The *group delay deviation* (GDD) is defined as the deviation in group delay of $\tau_x(k)$ with respect to τ_w (Stark and

---

3 For definitions of IFD and RPS and their details, see Chapter 2.
4 Further details are presented in Chapter 4.

Paliwal 2009),

$$\Delta \tau_x(k) = \tau_w - \tau_x(k). \tag{3.61}$$

As discussed in Chapter 2, the group delay deviation exhibits minima at spectral peaks (Stark and Paliwal 2009). Using this constraint along with the geometry, Mowlaee *et al.* (2012) proposed phase estimation for signal reconstruction in single-channel source separation where the group delay deviation based distance metric is given by

$$d_{a(k_h, l)} = 1 - \cos\left(\frac{2\pi}{N_w} \tau_w - (\hat{\phi}_x(k_h, l) - \hat{\phi}_x(k_h + 1, l))\right). \tag{3.62}$$

The minimum group delay deviation constraint around harmonic peaks helps to select the correct phase candidate, which is unknown due to the ambiguity in the sign difference between two spectra. Employing $d_{a(k_h, l)}$ in order to remove the ambiguity in the phase candidates, the optimal phase value for frequency k_h is given by

$$\hat{\phi}_x^*(k_h, l) = \underset{\hat{\phi}_x^{(a)}(k_h, l)}{\arg\min} \; d_{a(k_h, l)}, \tag{3.63}$$

where $\hat{\phi}_x^{(a)}(k_h, l)$ is the set of ambiguous clean phase estimates.

3.3.6 Least Squares (LS)

Chacon and Mowlaee (2014) formulated a least squares solution to estimate the clean phase spectrum from a given mixture signal. For simplicity and compactness of the mathematical notation, in the following we will drop l and k. Similar to (3.58) and from the general minimization problem in the complex domain,

$$d(\hat{\phi}_x) = |Y - (|X|e^{j\hat{\phi}_x} + D)|^2. \tag{3.64}$$

They further considered an auxiliary phase distortion cost function defined by a *mean cyclic error* (MCE) criterion given by

$$d_{MCE}(\phi_x, \hat{\phi}_x) = 1 - \cos(\phi_x - \hat{\phi}_x). \tag{3.65}$$

Figure 3.13 compares a squared error cost function $d_{MSE} = (\phi_x - \hat{\phi}_x)^2$ and the cyclic error cost function $d_{MCE} = 1 - \cos(\phi_x - \hat{\phi}_x)$ against the estimation error $-2\pi \le \phi_x - \hat{\phi}_x \le 2\pi$. The absolute error within $-2\pi \le \phi_x - \hat{\phi}_x \le 2\pi$ increases for the

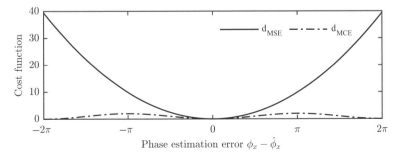

Figure 3.13 Comparison between phase estimation error criteria: squared error ($d_{MSE}(\phi_x, \hat{\phi}_x)$) and cyclic error ($d_{MCE}(\phi_x, \hat{\phi}_x)$) measures shown versus phase estimation error $\phi_x - \hat{\phi}_x$. For further details, see Nitzan *et al.* (2016).

squared-error cost function as it takes no periodicity into account. In contrast, the cyclic cost function decreases, reflecting the periodic nature of the cyclic random variable estimation problem. The cyclic error (Nitzan *et al.* 2016), in contrast to the mean square error (MSE) criterion, is invariant to modulo 2π in the estimation error. For small estimation errors, the cyclic error resembles the squared error, since we have: $1 - \cos(\phi_x - \hat{\phi}_x) \approx \frac{(\phi_x - \hat{\phi}_x)^2}{2}$ for $\phi_x - \hat{\phi}_x < 1$. For more details on Bayesian cyclic bounds and periodic parameter estimation using the mean cyclic error cost function, see Nitzan *et al.* (2016).

Solving the optimization problem in (3.64), together with the constraint (3.65), results in formulating a constrained optimization defined in the following. Applying Lagrangian multipliers, the constrained problem can be formulated as minimizing the Lagrangian function

$$\Lambda(\hat{\phi}_x, \lambda) = \|Y|e^{j\phi_y} - |X|e^{j\hat{\phi}_x} - |D|e^{j\phi_d}|^2 - \lambda(1 - \cos(\phi_x - \hat{\phi}_x)), \tag{3.66}$$

and the resulting LS phase estimate is given by

$$\hat{\phi}_x^{ls} = \arg\min_{\hat{\phi}_x} \Lambda(\hat{\phi}_x, \lambda). \tag{3.67}$$

Using (3.57) and the fact that $2\cos(x) = e^{jx} + e^{-jx}$, and finding $\hat{\phi}_x$ as the solution to $\nabla_{\hat{\phi}_x, \lambda} \Lambda = \mathbf{0}$, the *least squares phase estimation* (LSPE) is given by

$$\phi_x^{ls}(k, l) = \phi_y(k, l) + j \log\left(\frac{X(k, l) + D(k, l)e^{-j\Delta\phi(k, l)}}{Y(k, l)}\right). \tag{3.68}$$

To have a real-valued LSPE, (3.68) results in the following constraint:

$$\text{Re}\left(\log\left(\frac{|X(k, l)| + |D(k, l)|(\cos(\Delta\phi(k, l)) + j\sin(\Delta\phi(k, l)))}{|Y(k, l)|}\right)\right) = 0. \tag{3.69}$$

Let $z^c = ze^{j\phi_z}$. Using the fact that $\log(z^c) = \log(z) + j\phi_z$, (3.69) can be rewritten as

$$\frac{|D(k, l)| \sin(\Delta\phi(k, l))}{Y(k, l)} = \pm\sqrt{1 - \left(\frac{|X(k, l)| + |D(k, l)| \cos\Delta\phi(k, l)}{|Y(k, l)|}\right)^2}. \tag{3.70}$$

This condition is always met, and therefore $\phi_x^{ls} \in \mathbb{R}$ for all $\phi_y(k, l), |Y(k, l)|, |X(k, l)|$, and $|D(k, l)|$.

For very high SNRs, the LSPE simplifies to $\lim_{D\to 0} \phi_x^{ls} = \phi_y$, which is consistent with Vary (1985), explaining why phase estimation has been reported unimportant for signal components of large local SNRs. Similar to the discussion in Vary (1985), Chacon and Mowlaee proposed taking into account the fact that a threshold SNR, called SNR_{th}, exists above which the noisy phase suffices so that the resulting least squares phase estimator is

$$\hat{\phi}_x^{LS}(k, l) = \begin{cases} \hat{\phi}_x^{ls}(k, l) & \frac{|X(k, l)|^2}{|D(k, l)|^2} < \text{SNR}_{th}, \\ \phi_y(k, l) & \text{otherwise.} \end{cases} \tag{3.71}$$

Using a Monte Carlo simulation, the threshold SNR (SNR_{th}) was found to be equal to 6 dB, which is consistent with the previous findings (Vary 1985).

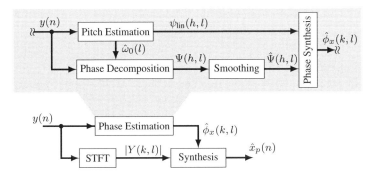

Figure 3.14 Temporal smoothing of unwrapped phase to estimate the clean phase from a noisy speech input. The steps are: fundamental frequency estimation, phase decomposition, temporal smoothing, and signal reconstruction.

3.3.7 Spectro-Temporal Smoothing of Unwrapped Phase

Kulmer *et al.* (2014) proposed applying a harmonic phase decomposition in order to obtain an unwrapped phase. The bottom line of their proposal is that the variance of the noisy phase spectrum is large due to the impact of noise compared to the clean phase spectrum. By reducing this large variance, a more consistent spectral pattern, closer to the clean phase, is achievable. The main idea and the steps involved are shown in the block diagram in Figure 3.14.

Kulmer and Mowlaee proposed the idea of smoothing the unwrapped phase for phase estimation of clean speech from its noisy observation (Kulmer *et al.* 2014; Kulmer and Mowlaee 2015a; Mowlaae and Kulmer 2015b). The common steps in all three contributions are summarized in the following steps: (i) pitch-synchronous segmentation for phase decomposition, (ii) linear phase removal to obtain the unwrapped phase estimate, (iii) applying a smoothing filter on the estimated unwrapped phase to find its enhanced version, (iv) using the enhanced, filtered unwrapped phase to reconstruct an enhanced-phase signal. Further details are provided in the following.

3.3.7.1 Signal Segmentation

The pitch-synchronous analysis described by Degottex and Erro, called *harmonic model phase distortion* (HMPD), is first applied. This step is required to get access to the individual phase components provided by the phase decomposition principle (see Chapter 2 for details). The time instants from the pitch-synchronous segmentation will be $t(l) = t(l-1) + \frac{1}{4f_0(l-1)}$, with $t(l)$ and $t(l-1)$ the time instants for the lth and $(l-1)$th frames, respectively, and $f_0(l)$ the fundamental frequency at the lth frame. At voiced frames, the underlying speech segment is modeled as a sum of harmonics with amplitude $|X(h,l)|$, and the instantaneous phase is $\psi(h,l)$, where h is defined as the harmonic index.

The instantaneous STFT phase sampled at harmonic h and frame l is decomposed into *linear* and *unwrapped parts*:

$$\psi(h,l) = \underbrace{h \sum_{l'=0}^{l} \omega_0(l')(t(l') - t(l'-1))}_{\text{Linear phase: } \psi_{\text{lin}}(h,l)} + \underbrace{\angle V(h,l) + \psi_d(h,l)}_{\text{Unwrapped phase: } \Psi(h,l)}, \tag{3.72}$$

where $V(h, l)$ is the vocal tract filter, $\Psi(h, l)$ is the unwrapped phase, and $\psi_{\text{lin}}(h, l)$ is the linear phase. The linear phase component wraps the instantaneous STFT phase.[5]

3.3.7.2 Linear Phase Removal

Discontinuities in the linear phase result in degradation of the perceived speech quality of the reconstructed speech signal (Stylianou 2001). Let $\hat{\omega}_0(l) = 2\pi \hat{f}_0(l)/f_s$ be the fundamental frequency estimate at frame l, where $\hat{f}_0(l)$ and f_s denote the fundamental frequency estimate and the sampling frequency, respectively. Then the linear phase is approximated as

$$\hat{\psi}_{\text{lin}}(h, l) = h \sum_{l'=0}^{l} \hat{\omega}_0(l')(t(l') - t(l' - 1)). \tag{3.73}$$

The unwrapped phase is calculated by subtracting the linear phase part from the instantaneous phase, and is given by

$$\Psi(h, l) = \psi(h, l) - \hat{\psi}_{\text{lin}}(h, l). \tag{3.74}$$

3.3.7.3 Apply Smoothing Filter

Three proposals were presented: temporal smoothing of unwrapped phase (Kulmer and Mowlaee 2015a); selective threshold-based time–frequency smoothing (Kulmer *et al.* 2014); and SNR-based time–frequency smoothing (Mowlaae and Kulmer 2015b).

3.3.7.3.1 Temporal Smoothing

A temporal smoothing filter was applied on the estimated unwrapped phase $\Psi(h, l)$ in the circular domain according to

$$\hat{\Psi}(h, l) = \angle \frac{1}{|\mathcal{R}|} \sum_{l' \in \mathcal{R}} e^{j\Psi(h, l')}, \tag{3.75}$$

where $\mathcal{R}$ is the set of neighboring frames lying within a time span of 20 ms. The authors applied the smoothing filter in regions with a high voicing probability (Kulmer and Mowlaee 2015a). Kulmer *et al.* (2014) proposed to selectively smooth using a probabilistic approach relying on a pre-defined threshold $\sigma_{c,\text{th}}^2$ on the circular variance of the unwrapped phase of each harmonic. The estimated $\hat{\Psi}(h, l)$ is obtained by the estimation of the circular mean if the circular variance is below the threshold (Kulmer *et al.* 2014):

$$\hat{\Psi}(h, l) = \begin{cases} \angle \frac{1}{|\mathcal{R}|} \sum_{l' \in \mathcal{R}} e^{j\Psi(h, l')}, & \text{if } \sigma_c^2(h, l) < \sigma_{c,\text{th}}^2 \\ \Psi(h, l), & \text{otherwise.} \end{cases} \tag{3.76}$$

Finally, the enhanced harmonic phase $\hat{\psi}(h, l)$ is obtained by adding the linear phase $\hat{\psi}_{\text{lin}}(h, l)$ back to the temporally smoothed unwrapped phase $\hat{\Psi}(h, l)$ given in (3.75), and we have

$$\hat{\psi}(h, l) = \hat{\Psi}(h, l) + \hat{\psi}_{\text{lin}}(h, l). \tag{3.77}$$

---

5 See Chapter 2 for further details on phase unwrapping and *harmonic model phase distortion* (HMPD).

3.3.7.3.2 Spectral Smoothing From the relative phase shift representation for unwrapped phase we define

$$R\hat{P}S_\psi(h, l) = \hat{\psi}(h, l) - h\hat{\psi}(1, l). \tag{3.78}$$

Mowlaee and Kulmer proposed frequency smoothing by applying the spectral smoothing filter to obtain an enhanced spectrally smoothed RPS denoted by $\overline{RPS}_{\hat{\psi}}(h, l)$ given by (Mowlaee and Kulmer 2015b)

$$\hat{\hat{\psi}}(h, l) = \overline{RPS}_{\hat{\psi}}(h, l) + h\hat{\psi}(1, l), \tag{3.79}$$

where $\hat{\hat{\psi}}(h, l)$ denotes the final time–frequency smoothed harmonic phase and $\hat{\psi}(1, l)$ is given by (3.77).

3.3.7.3.3 SNR-Based Smoothing As reported in Gerkmann (2014), the concentration parameter for phase is SNR dependent. Therefore, Mowlaee and Kulmer (2015b) formulated the phase smoothing idea in a SNR-based framework expressed as a binary hypothesis test that tackles the phase estimation problem in joint detection–estimation. They used the concentration parameter in the von Mises distribution controlled by the SNR information, as explained below.

Considering a two class problem definition for phase estimation, we have

$$H_0 : \hat{\Psi}_x(h, l) = \Psi(h, l) \tag{3.80}$$

$$H_1 : \hat{\Psi}_x(h, l) = \Psi(h, l) + e(h, l), \tag{3.81}$$

where $\Psi_x(h, l)$ and $\Psi(h, l)$ are the unwrapped phase components for the clean and noisy speech signals, respectively, and $e(h, l)$ is the error term that captures the remaining uncertainty in the phase estimate. The hypothesis H_0 denotes the case of no harmonic structure in the spectral phase. Hence, the spectral phase can be assumed to be distributed uniformly within the range of the phase variable $[-\pi, \pi[$, and we have

$$p(\Psi(h, l)|H_0) = \frac{1}{2\pi}. \tag{3.82}$$

The uniform distribution reveals no information about a harmonic structure in phase (hence maximum uncertainty). On the other hand, hypothesis H_1 denotes the case of a harmonic excitation in which the spectral phase can be modeled using a von Mises distribution given by

$$p(\Psi(h, l)|H_1) = \frac{e^{\kappa(h,l)\cos(\Psi(h,l)-\Psi_\mu(h,l))}}{2\pi I_0(\kappa(h, l))}, \tag{3.83}$$

with $\Psi_\mu(h, l)$ the mean and $\kappa(h, l)$ the concentration parameter to statistically model the error $e(h, l)$. Finally, given these two hypotheses, the decision to accept either of them is given by (Kay 1998)

$$R(H_0 : H_1) = \frac{p(\Psi(h, l)|H_0)}{p(\Psi(h, l)|H_1)}$$

$$= \frac{I_0(\kappa(h, l))}{e^{\kappa(h,l)\cos(\Psi(h,l)-\Psi_\mu(h,l))}}. \tag{3.84}$$

Applying the decision rule as the binary hypothesis test, $R(H_0 : H_1) \overset{H_0}{\underset{H_1}{\gtrless}} 1$, and taking the natural logarithm from both sides, we finally obtain

$$\cos(\Psi(h,l) - \Psi_\mu(h,l)) \overset{H_1}{\underset{H_0}{\gtrless}} \theta_{\text{th}}(h,l) = \frac{\ln I_0(\kappa(h,l))}{\kappa(h,l)}. \tag{3.85}$$

Therefore, in the following, we present how to assign the uncertainty in phase captured by $\kappa(h,l)$ using the SNR information. Mowlaee and Kulmer proposed to find an SNR-based estimate for $\cos(\Psi_{\text{dev}}(h,l)) = \cos(\Psi(h,l) - \Psi_\mu(h,l))$,

$$\cos\Psi_{\text{dev}}(h,l) = \frac{\text{SNR}_{\text{prior}}(h,l) + \text{SNR}_{\text{post}}(h,l) - 1}{2\sqrt{\text{SNR}_{\text{post}}(h,l)\text{SNR}_{\text{prior}}(h,l)}}, \tag{3.86}$$

defining the local *a posteriori* SNR as $\text{SNR}_{\text{post}}(h,l) = \frac{|Y(h,l)|^2}{|D(h,l)|^2}$ and local *a priori* SNR as $\text{SNR}_{\text{prior}}(h,l) = \frac{|X(h,l)|^2}{|D(h,l)|^2}$. Given the estimates for prior and posterior SNRs, an estimate for $\cos\Psi_{\text{dev}}(h,l)$ can be obtained by using (3.86). The right-hand side of (3.85) is then obtained by calculating $\kappa(h,l)$ for every harmonic and frame:

$$\sigma_c^2(h,l) = 1 - \frac{I_1(\kappa(h,l))}{I_0(\kappa(h,l))}. \tag{3.87}$$

The estimated cosine phase deviation is then compared to the threshold $\theta_{\text{th}}(h,l)$ for the decision-making in the binary hypothesis test. The tempo-spectral phase smoothing is then performed only when the phase deviation cosine exceeds the threshold $\theta_{\text{th}}(h,l)$ in (3.85) controlled by the concentration parameter $\kappa(h,l)$,

$$\hat{\Psi}_x(h,l) = \begin{cases} \angle\frac{1}{|\mathcal{R}|}\sum_{l'\in\mathcal{R}} e^{i\Psi(h,l')}, & \text{if } \cos(\Psi_{\text{dev}}(h,l)) \geq \theta_{\text{th}}(h,l), \\ \Psi(h,l), & \text{otherwise.} \end{cases} \tag{3.88}$$

Figure 3.15 shows the regions for the two hypotheses H_0 and H_1 depending on the estimated concentration parameter $\kappa(h,l)$ and $\cos(\Psi_{\text{dev}}(h,l))$.

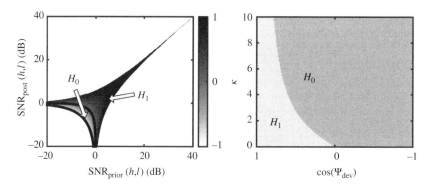

Figure 3.15 SNR-based smoothing (Mowlaee and Kulmer 2015b): Phase deviation cosine $-1 \leq \cos(\Psi_{\text{dev}}) \leq 1$ as a function of *a priori* and *a posteriori* local SNRs (left); the regions for hypotheses H_0 and H_1 depending on the values of κ and $\cos(\Psi_{\text{dev}})$ (right).

3.3.7.4 Reconstruction of the Enhanced-Phase Signal

The smoothing of unwrapped phase methods described above is carried out at harmonics, while for signal reconstruction and to be capable of combining with an amplitude-only speech enhancement method, an enhanced STFT phase is required. Therefore, the estimated harmonic phase $\hat{\psi}(h,l) = \hat{\psi}_{\text{lin}}(h,l) + \hat{\Psi}_x(h,l)$ is transformed to the STFT domain by modifying the frequency bins within the mainlobe width of the analysis window denoted by $N_p = \Delta\omega_{\text{MW}} N_{\text{DFT}}/(2\pi)$,

$$\hat{\phi}(\lfloor h\omega_0 N_{\text{DFT}}/(2\pi)\rfloor + i,l) = \hat{\psi}(h,l), \ \forall i \in [-N_p/2, N_p/2], \tag{3.89}$$

where N_{DFT} is the DFT size. Using the estimated STFT phase $\hat{\phi}(k,l)$ and applying an overlap-add routine, the phase-enhanced signal $\hat{x}_p(n)$ is reconstructed.

3.4 Experiments

In this section, we provide four experiments in order to demonstrate how to estimate the clean phase spectrum from its noisy observation. The experiments are:

- Monte Carlo simulation to compare ML and MAP
- Monte Carlo simulation to illustrate the window impact on phase estimation
- Phase recovery using Griffin and Lim based methods
- Phase estimation for signal enhancement: a comparative study.

The first two experiments elaborate on the behavior and performance of the statistical phase estimators for synthetic data, leading to important conclusions with regards to phase estimation error dependence on the window length, type, and input SNR. The last two experiments pave the way to extend the phase estimators for practical speech processing applications.

3.4.1 Experiment 3.1: Monte Carlo Simulation Comparing ML and MAP

In the following, we present a performance evaluation of ML and MAP phase estimators to estimate phase values from an artificially generated signal observed in noise. The signal consists of harmonics up to the Nyquist frequency of $f_s/2$ and additive white Gaussian noise. The sampling frequency is set to $f_s = 8$ kHz. The amplitudes of the harmonics are equally set to $A_h = 1$. We perform 10^5 Monte Carlo runs with uniformly distributed phase values $\phi_h \sim \mathcal{U}[-\pi, \pi[$. The performance is evaluated using the *cyclic mean phase error*

$$d_{\text{MCE}} = \frac{1}{H}\sum_{h=1}^{H}\overline{d}_{\text{MCE}}(\phi_h, \hat{\phi}_h), \tag{3.90}$$

where $\overline{d}_{\text{MCE}}(\phi_h, \hat{\phi}_h) = (1 - \cos(\phi_h - \hat{\phi}_h))$, and ϕ_h and $\hat{\phi}_h$ are the correct and the estimated phase at harmonic h, respectively. Figure 3.16 illustrates the impact of additive noise on the ML (3.33) and MAP phase estimates (3.41) depending on the frequencies $\omega_0 \in [0.05, 0.25]$. Since $\omega_0 = 2\pi f_0/f_s$, the frequency range of ω_0 corresponds to a speech processing relevant fundamental frequency range of $f_0 \in [62.5, 312.5]$ Hz at $f_s = 8$ kHz.

The ML phase estimation in Figure 3.16 shows a strong dependency on ω_0, especially for high SNRs. It is interesting to note that the phase error is non-zero even for

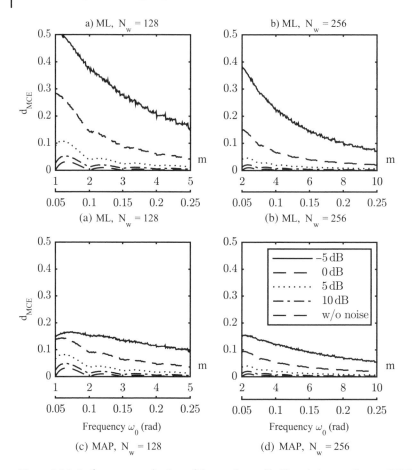

Figure 3.16 Performance evaluation of the maximum likelihood phase estimator (3.33) and maximum *a posteriori* estimator (3.41) with regard to fundamental frequency, signal-to-noise ratio, and a window length of $N_w = 128$ [(a),(c)] and $N_w = 256$ [(b),(d)]. The non-zero phase error for low noise scenarios is caused by the approximation in (3.31).

the noise-free scenario where the phase error decreases for higher frequencies ω_0 and becomes zero if m approaches an integer. This effect is caused by the approximation made in (3.31) where the terms on the right-hand side of (3.27) were neglected. Low frequencies violate the assumption of ω_0 not close to 0 or π and lead to a biased phase estimate. In the case that $m = \omega_0 \frac{N_w}{2\pi}$ approaches an integer, the observation window consists of full periods of sinusoids only, the spectral leakage effect is minimized, and the phase error becomes zero.

The dependency on the window length is shown for $N_w = 128$ (a) and $N_w = 256$ (b). A longer observation interval of $N_w = 256$ decreases the error due to more available measurement points and less impact of the window length dependent interfering term in (3.30).

The MAP phase estimator benefits from the prior information of a phase distribution. This information is obtained from previous observations of the noisy signal under the assumption of weak stationarity. In case of speech processing the considered signal could represent a phoneme, which are often assumed to be stationary within a range of 100 ms. Then the parameters of the prior phase distribution for the current frame can be estimated using the previous frames. For the performance evaluation we set the parameters of the von Mises distribution as follows: the mean μ_h is equal to the true phase value $\mu_h = \phi_h$ and the concentration parameter $\kappa_h = 1$ for all harmonics h. Figure 3.16(c) and (d) illustrate the performance of the MAP estimator with different SNRs. Compared to the ML estimator, the MAP is more robust to additive noise for both window lengths of $N_w = 128$ (c) and $N_w = 256$ (d), as it gains prior information about the phase values. However, even for the noise-free scenario the phase error does not approach zero due to the approximations made in (3.39).

Finally, we can conclude that both ML and MAP estimators suffer from spectral leakage. The phase error approaches zero only if the window length is a multiple of the fundamental frequency. Further, a longer window length is beneficial under low SNR conditions. Please note that both estimators were derived using a rectangular window, which has rather poor sidelobe rejection. Considering other window types, for example Hann or Hamming, the phase error can be further decreased. The MATLAB® implementation for this experiment can be found in *Exp3_1.m* in the *PhaseLab Toolbox* (see the appendix).

3.4.2 Experiment 3.2: Monte Carlo Simulation on Window Impact

This experiment shows the impact of different window functions on the phase estimation. We again consider a signal $x(n)$ consisting of harmonics h up to the Nyquist frequency of $f_s/2$ with equal amplitudes $A_h = 1$. We want to estimate the phase values under noisy conditions of additive, white Gaussian noise $d(n)$ with an SNR of 0 dB by using the discrete-time Fourier transform

$$x(n) = \sum_{h=1}^{H} A_h \cos(h\omega_0 + \phi_h) + d(n), \tag{3.91}$$

$$X_w(e^{j\omega}) = \text{DTFT}(x(n)w(n)), \tag{3.92}$$

$$\hat{\phi}_h = \angle X_w(e^{jh\omega_0}), \tag{3.93}$$

with $w(n)$ denoting the window function. Figure 3.17 presents the cyclic mean phase error defined in (3.90). In this experiment, we consider a frequency range $\omega_0 = 2\pi\frac{f_0}{f_s}$ of $\omega_0 \in [0.05, 0.25]$. The performance is evaluated for different window lengths of $N_w = 127$ (left) and $N_w = 255$ (right). The symmetric window functions with $n \in [-(N_w - 1)/2, (N_w - 1)/2]$ fulfill the zero phase property (see Figure 3.17) which is beneficial in the case of inaccurate fundamental frequency estimates. We evaluate the phase error for three different cases: fundamental frequency is known (top row), and fundamental frequency is underestimated by 2.5% (middle) and 5% (bottom).

In case of a known fundamental frequency, the rectangular window performs best. Its high frequency resolution due to the small mainlobe width suppresses the surrounding additive noise. The wider mainlobe width of the Blackman window contains more noise power, which results in a lower performance in general. Further, a wider mainlobe width hinders the separation of the harmonics, especially at low frequencies.

In case of an underestimated fundamental frequency estimate of 2.5% and 5% the phase errors are in general higher, showing the importance of an accurate fundamental

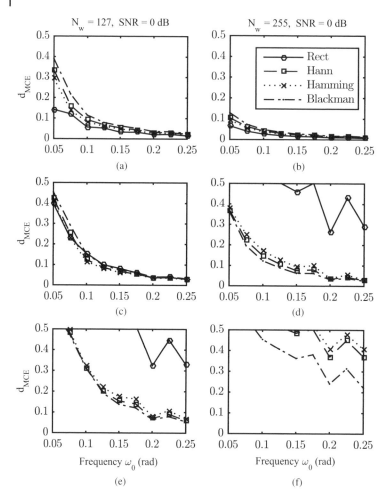

Figure 3.17 Impact of different window functions on phase estimation for an accurate fundamental frequency f_0 estimate (top row), and underestimated f_0 by 2.5% (middle) and 5% (bottom) for a window length of $N_w = 127$ (left) and $N_w = 255$ (right), revealing the importance of an accurate f_0 estimate. Given an accurate f_0 (top), the rectangular window performs best due to its high frequency resolution. For inaccurate f_0 estimates (bottom), window functions with wider mainlobes, e.g. the Blackman window, are in favor.

frequency estimation. The rectangular window suffers strongly as the underestimated fundamental frequency does not lie within the mainlobe width. Window functions with a broader mainlobe width are more robust to biased frequency estimates as indicated in Figure 3.17(f), where the Blackman window with the widest mainlobe performs best.

We can conclude that selecting a window function depends strongly on the application. If a highly accurate fundamental frequency estimate is given then the rectangular window performs best due to its high frequency resolution. In the case of an inaccurate fundamental frequency estimate, a window function with a wider mainlobe width (e.g., Blackman) is beneficial for phase estimation. The MATLAB® implementation for this experiment can be found in *Exp3_2.m* in the *PhaseLab Toolbox* (see the appendix).

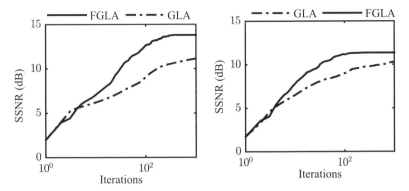

Figure 3.18 Griffin and Lim algorithm (GLA; Griffin and Lim 1984: dashed) and fast Griffin and Lim algorithm (solid; Perraudin *et al.* 2013) used for phase recovery. The results are shown in SSNR measured in dB for (left) female and (right) male speech versus the number of iterations.

3.4.3 Experiment 3.3: Phase Recovery Using the Griffin–Lim Algorithm

We study the performance of the Griffin–Lim algorithm (Griffin and Lim 1984) and its extended version called the *fast Griffin–Lim algorithm* (FGLA) proposed in Perraudin *et al.* (2013). We provide examples for signal estimation from its spectrogram (phase reconstruction). The goal is to find a signal given its spectrogram. As the initial phase, zero values are used. The number of iterations in the Griffin–Lim algorithm is set to 10 000. The MATLAB® implementation for this experiment can be found in *Exp3_3.m* in the *PhaseLab Toolbox*. The MATLAB® package *The Large Time–Frequency Analysis Toolbox* (LTFAT) is needed to run this experiment[6] (for more details see Průša *et al.* (2014)).

As an evaluation criterion, the SSNR measure in dB is used to monitor the speed of convergence, defined as follows:

$$\text{SSNR}(\mathbf{x}) = -10 \log_{10} \frac{\|\mathbf{x}\|_2}{\|\hat{\mathbf{x}} - \mathbf{x}\|_2}, \tag{3.94}$$

where $\mathbf{x} = [x(0), \dots, x(N-1)]$ is the original desired signal and $\hat{\mathbf{x}}$ the reconstructed signal using GLA/FGLA. The results are shown in Figure 3.18 for (left) female and (right) male speech. It can be seen that FGLA converges faster and results in a larger averaged SSNR.

3.4.4 Experiment 3.4: Phase Estimation for Speech Enhancement: A Comparative Study

Figure 3.19 shows the block diagram for single-channel speech enhancement where the noisy signal is often enhanced in two steps: (i) amplitude estimation, and (ii) signal reconstruction. In Chapter 4, we will see how an estimated clean phase may contribute to finding a more accurate spectral amplitude estimate. In this experiment, we only focus on signal reconstruction where the noisy phase spectrum is replaced by an estimated clean phase using either of the phase estimators discussed in this chapter. The experiment demonstrates how an estimated clean phase spectrum positively impacts the signal reconstruction in single-channel speech enhancement.

---

6 Downloadable from http://ltfat.github.io/phaseret/mat/gla

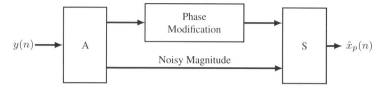

Figure 3.19 Block diagram for the single-channel speech enhancement example used in Experiment 3.4 to demonstrate the effectiveness of phase estimators when used to replace the noisy spectra phase at signal reconstruction. Phase modification refers to any selected phase estimator listed in Table 3.1; "A" and "S" denote the analysis and synthesis steps.

We quantify the improvement brought by replacing the noisy phase with the estimated clean phase obtained by each method. No explicit amplitude enhancement scheme is used; therefore, any reported improvement here should be considered due to the phase-only enhancement. The results are shown for the MAP estimator (Kulmer and Mowlaee 2015b), phase randomization (Sugiyama and Miyahara 2013), model-based phase reconstruction (STFTPI) (Krawczyk and Gerkmann 2014), geometry-based (Mowlaee *et al.* 2012), and *temporal smoothing of unwrapped phase* (TSUP) (Kulmer *et al.* 2014). Results for clean phase and noisy phase as the upper and lower bounds are also included for comparison purposes.

The results are shown in Figure 3.20 for different signal-to-noise ratios in terms of PESQ (Rix *et al.* 2001) and STOI (Taal *et al.* 2011), commonly used in speech enhancement literature to quantify the perceived quality and speech intelligibility, respectively.[7] As will be discussed in Chapter 6, some artifacts due to phase harmonization lead to an improved PESQ score but at the expense of a buzzy speech quality. To circumvent this issue, and in order to quantify the phase estimation accuracy, we further employ the unwrapped root mean square estimation error (UnRMSE) in dB defined as:

$$\text{UnRMSE} = 10 \log_{10}\left(\frac{1}{L} \sum_{l=1}^{L} \sqrt{\frac{\sum_{h=1}^{H} |X(h,l)|^2 (\Psi_x(h,l) - \hat{\Psi}_x(h,l))^2}{\sum_{h}^{H} |X(h,l)|^2}} \right), \tag{3.95}$$

where $X(h,l)$ is the spectral amplitude sampled at harmonic h and frame l, and $\hat{\Psi}_x(h,l)$ and $\Psi_x(h,l)$ are the corresponding unwrapped harmonic phase spectra. UnRMSE reveals how close an estimated phase is to its clean phase reference, meaning that the lower it is, the more accurate the applied phase estimator. The MATLAB® implementation for this experiment can be found in *Exp3_4.m* in the *PhaseLab Toolbox* (for further details, see the appendix).

The following observations can be made:

- Compared to other phase estimators, STFTPI provides slightly better PESQ improvement in babble noise at the expense of degraded intelligibility, lower than the unprocessed signal. This is due to the buzzyness in the reconstructed signal (as reported in Patil and Gowdy 2014, Krawczyk and Gerkmann 2014), with a large UnRMSE score.
- Phase randomization achieves a comparable PESQ score to STFTPI, in particular for the stationary noise scenario, however at the expense of worse performance in terms of speech intelligibility (STOI) and phase estimation accuracy (UnRMSE).

---

7 For a detailed review of speech quality estimation, see Chapter 6

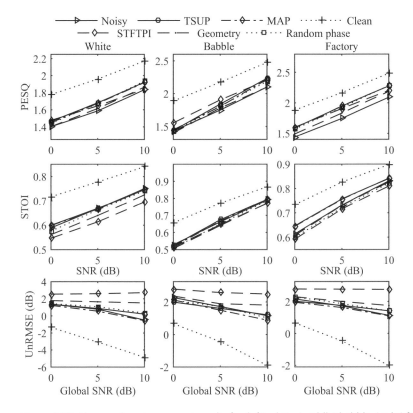

Figure 3.20 Phase-only enhancement results for (left) white, (middle) babble, (right) factory noise, reported in PESQ (top row), STOI (middle), and unRMSE (bottom).

- The MAP phase estimator provides a good trade-off between joint improvement in quality and intelligibility, predicted by PESQ and STOI, respectively. It results in the lowest UnRMSE, close to that of the clean phase as the phase estimation upper bound. MAP slightly outperforms TSUP.

3.5 Summary

This chapter presented the phase estimation problem and provided a detailed explanation on how it is tackled using signal processing tools. We started with the simple scenario of estimating the phase of one sinusoid and extended it to multiple sinusoids in additive noise. Several phase estimation methods were presented and discussed, including maximum likelihood, maximum *a posteriori*, least squares, geometry-based, and model-based approaches. Detailed analysis revealed the limits and potential of the existing statistical phase estimators used to resolve the phase of one or multiple sinusoids in noise. The impact of the window used in the spectral analysis on the phase estimation error mean and variance was explained. This led to some guidelines to reduce the cyclic error or the error variance in the phase estimation problem.

A full literature review on the existing solutions proposed to estimate the clean phase spectrum in the context of speech enhancement was presented. The principle used in each method and some proof-of-concepts have been provided to present an insightful review of the phase estimation topic. Through experiments, we demonstrated the usefulness of the different phase estimators discussed in this chapter. Comparative studies were carried out as Monte Carlo simulations on synthetic data as well as tests on realistic noisy speech in order to demonstrate the usefulness of the available phase estimators in the context of the application of phase retrieval from spectral magnitude or phase estimation for improved speech enhancement.

References

J. B. Allen, Short term spectral analysis, synthesis, and modification by discrete Fourier transform, *IEEE Transactions on Acoustics, Speech, and Signal Processing*, vol. 25, no. 3, pp. 235–238, 1977.

L. D. Alsteris and K. K. Paliwal, Iterative reconstruction of speech from short-time Fourier transform phase and magnitude spectra, *Computer Speech & Language*, vol. 21, no. 1, pp. 174–186, 2007.

C. Chacon and P. Mowlaee, *Least Squares Phase Estimation of Mixed Signals*, Proceedings of the International Conference on Spoken Language Processing (INTERSPEECH), pp. 2705–2709, 2014.

G. Degottex and D. Erro, A uniform phase representation for the harmonic model in speech synthesis applications, *EURASIP Journal on Audio, Speech, and Music Processing*, no. 1, pp. 1–16, 2014.

M. Dolson, The phase vocoder: A tutorial, *Computer Music Journal*, vol. 10, no. 4, pp. 14–27, 1986.

T. Gerkmann, Bayesian estimation of clean speech spectral coefficients given *a priori* knowledge of the phase, *IEEE Transactions on Signal Processing*, vol. 62, no. 16, pp. 4199–4208, 2014.

D. Griffin and J. Lim, Signal estimation from modified short-time Fourier transform, *IEEE Transactions on Acoustics, Speech, and Signal Processing*, vol. 32, no. 2, pp. 236–243, 1984.

F. J. Harris, On the use of windows for harmonic analysis with the discrete Fourier transform, *Proceedings of the IEEE*, vol. 66, no. 1, pp. 51–83, 1978.

M. Hayes, J. Lim, and A. V. Oppenheim, Signal reconstruction from phase or magnitude, *IEEE Transactions on Acoustics, Speech, and Signal Processing*, vol. 28, no. 6, pp. 672–680, 1980.

R. C. Hendriks, T. Gerkmann, and J. Jensen, *DFT-Domain-Based Single-Microphone Noise Reduction for Speech Enhancement*, Morgan & Claypool Publishers, Synthesis Lectures on Speech and Audio Processing, 2013.

S. M. Kay, *Fundamentals of Statistical Signal Processing, Volume I: Estimation Theory*, Prentice Hall, 1993.

S. M. Kay, *Fundamentals of Statistical Signal Processing, Volume II: Detection Theory*, Prentice Hall, 1998.

T. Kleinschmidt, S. Sridharan, and M. Mason, The use of phase in complex spectrum subtraction for robust speech recognition, *Computer Speech & Language*, vol. 25, no. 3, pp. 585–600, 2011.

M. Krawczyk and T. Gerkmann, STFT phase reconstruction in voiced speech for an improved single-channel speech enhancement, *IEEE/ACM Transactions on Audio, Speech, and Language Processing*, vol. 22, no. 12, pp. 1931–1940, 2014.

J. Kulmer, P. Mowlaee, and M. K. Watanabe, *A Probabilistic Approach For Phase Estimation in Single-Channel Speech Enhancement Using von Mises Phase Priors*, Proc. IEEE International Workshop on Machine Learning for Signal Processing (MLSP), 2014.

J. Kulmer and P. Mowlaee, Phase estimation in single channel speech enhancement using phase decomposition, *IEEE Signal Processing Letters*, vol. 22, no. 5, pp. 598–602, 2015a.

J. Kulmer and P. Mowlaee, *Harmonic Phase Estimation in Single-Channel Speech Enhancement Using von Mises Distribution and Prior SNR*, Proceedings of the IEEE International Conference on Acoustics, Speech and Signal Processing (ICASSP), pp. 5063–5067, 2015b.

J. Le Roux, E. Vincent, Y. Mizuno, H. Kameoka, N. Ono, and S. Sagayama, *Consistent Wiener Filtering: Generalized Time-Frequency Masking Respecting Spectrogram Consistency*, Proceedings of the International Conference on Latent Variable Analysis and Signal Separation, pp. 89–96, 2010a.

J. Le Roux, H. Kameoka, N. Ono, and S. Sagayama, *Fast Signal Reconstruction from Magnitude STFT Spectrogram based on Spectrogram Consistency*, Proceedings of the International Conference on Digital Audio Effects (DAFx), 2010b.

R. McGowan and R. Kuc, A direct relation between a signal time series and its unwrapped phase, *IEEE Transactions on Acoustics, Speech and Signal Processing*, vol. 30, no. 5, pp. 719–726, 1982.

K. V. Mardia and P. E. Jupp, *Directional Statistics*, John Wiley & Sons, 2008.

E. Mehmetcik and T. Ciloglu, Speech enhancement by maintaining phase continuity between consecutive analysis frames, *The Journal of the Acoustical Society of America*, vol. 132, no. 3, 2012.

R. Miyahara and A. Sugiyama, *An Auto-Focusing Noise Suppressor for Cellphone Movies Based on Peak Preservation and Phase Randomization*, Proceedings of the IEEE International Conference on Acoustics, Speech and Signal Processing (ICASSP), pp. 2785–2789, 2013.

R. Miyahara and A. Sugiyama, *An Auto-Focusing Noise Suppressor for Cellphone Movies Based on Phase Randomization and Power Compensation*, Proceedings of the IEEE International Conference on Acoustics, Speech and Signal Processing (ICASSP), pp. 2199–2203, 2014.

P. Mowlaee and J. Kulmer, Phase estimation in single-channel speech enhancement: Limits-potential, *IEEE/ACM Transactions on Audio, Speech, and Language Processing*, vol. 23, no. 8, pp. 1283–1294, 2015a.

P. Mowlaee and J. Kulmer, Harmonic phase estimation in single-channel speech enhancement using phase decomposition and SNR information, *IEEE/ACM Transactions on Audio, Speech, and Language Processing*, vol. 23, no. 9, pp. 1521–1532, 2015b.

P. Mowlaee, R. Saeidi, and R. Martin, *Phase Estimation for Signal Reconstruction in Single-Channel Speech Separation*, Proceedings of the International Conference on Spoken Language Processing (INTERSPEECH), pp. 1548–1551, 2012.

P. Mowlaee and R. Saeidi, Iterative closed-loop phase-aware single-channel speech enhancement, *IEEE Signal Processing Letters*, vol. 20, no. 12, pp. 1235–1239, 2013.

P. Mowlaee and R. Saeidi, *Time-Frequency Constraint for Phase Estimation in Single-Channel Speech Enhancement*, Proceedings of The International Workshop on Acoustic Signal Enhancement (IWAENC), pp. 338–342, 2014.

S. Nawab, T. F. Quatieri, and J. S. Lim, *Algorithms for Signal Reconstruction from Short-Time Fourier Transform Magnitude*, Proceedings of the IEEE International Conference on Acoustics, Speech, and Signal Processing (ICASSP), pp. 800–803, 1983.

E. Nitzan, T. Routtenberg, and J. Tabrikian, A new class of Bayesian cyclic bounds for periodic parameter estimation, *IEEE Transactions on Signal Processing*, vol. 64, no. 1, pp. 229–243, 2016.

S. P. Patil and J. N. Gowdy, *Exploiting the Baseband Phase Structure of the Voiced Speech for Speech Enhancement*, Proceedings of the IEEE International Conference on Acoustics, Speech, and Signal Processing (ICASSP), pp. 6092–6096, 2014.

S. P. Patil and J. N. Gowdy, Use of baseband phase structure to improve the performance of current speech enhancement algorithms, *Speech Communication*, vol. 67, pp. 78–91, 2015.

N. Perraudin, P. Balazs, and P. L. Sondergaard, *A Fast Griffin–Lim Algorithm*, Proc. IEEE Workshop on Applications of Signal Processing to Audio and Acoustics (WASPAA), 2013.

K. Peters, S. Kay, *Unbiased Estimation of the Phase of a Sinusoid*, Proceedings of the IEEE International Conference on Acoustics, Speech, and Signal Processing, pp. 493–496, 2004.

Z. Průša, P. L. Søndergaard, N. Holighaus, C. Wiesmeyr, and P. Balazs, *The Large Time-Frequency Analysis Toolbox 2.0. Sound, Music, and Motion*, Lecture Notes in Computer Science, pp 419–442, 2014.

T. F. Quatieri and A. V. Oppenheim, Iterative techniques for minimum phase signal reconstruction from phase or magnitude, *IEEE Transactions on Acoustics, Speech and Signal Processing*, vol. 29, no. 6, pp. 1187–1193, 1981.

L. Rabiner and R. Schafer, *Digital Processing of Speech Signals*, Prentice Hall, 1978.

A. W. Rix, J. G. Beerends, M. P. Hollier, and A. P. Hekstra, *Perceptual Evaluation of Speech Quality (PESQ): A New Method for Speech Quality Assessment of Telephone Networks and Codecs*, Proceedings of the IEEE International Conference on Acoustics, Speech and Signal Processing (ICASSP), pp. 749–752, 2001.

P. A. Stark and K. K. Paliwal, *Group-Delay-Deviation Based Spectral Analysis of Speech*, Proceedings of the International Conference on Spoken Language Processing (INTERSPEECH), pp. 1083–1086, 2009.

P. Stoica and R. L. Moses, *Spectral Analysis of Signals*, Prentice Hall, 2005.

N. Sturmel and L. Daudet, *Signal Reconstruction from STFT Magnitude: A State of the Art*, Proc. International Conference on Digital Audio Effects DAFx, pp. 375–386, 2011.

Y. Stylianou, Applying the harmonic plus noise model in concatenative speech synthesis, *IEEE Transactions on Audio, Speech, and Language Processing*, vol. 9, no. 1, pp. 21–29, 2001.

A. Sugiyama and R. Miyahara, *Phase Randomization: A New Paradigm for Single-Channel Signal Enhancement*, Proceedings of the IEEE International Conference on Acoustics, Speech and Signal Processing (ICASSP), pp. 7487–7491, 2013.

C. H. Taal, R. C. Hendriks, R. Heusdens, and J. Jensen, An algorithm for intelligibility prediction of time-frequency weighted noisy speech, *IEEE/ACM Transactions on Audio, Speech, and Language Processing*, vol. 19, no. 7, pp. 2125–2136, 2011.

H. L. Van Trees, *Detection, Estimation, and Modulation Theory, Part 1*, John Wiley & Sons, 1968.

P. Vary, Noise suppression by spectral magnitude estimation mechanism and theoretical limits, *Signal Processing*, vol. 8, no. 4, pp. 387–400, 1985.

X. Zhu, G. T. Beauregard, and L. Wyse, *Real-Time Iterative Spectrum Inversion with Look-Ahead*, Proceedings of the IEEE International Conference on Multimedia and Expo (ICME), pp. 229–232, 2006.

X. Zhu, G. Beauregard, and L. L. Wyse, Real-time signal estimation from modified short-time Fourier transform magnitude spectra, *IEEE Transactions on Audio, Speech, and Language Processing*, vol. 15, no. 5, pp. 1645–1653, 2007.

Part II

Applications

4

Phase Processing for Single-Channel Speech Enhancement

Johannes Stahl and Pejman Mowlaee

Graz University of Technology, Graz, Austria

4.1 Introduction and Chapter Organization

The previous chapters have given an introduction on how to tackle the phase estimation problem in general and for speech processing in particular. In this chapter we will consider the knowledge gained in the context of speech enhancement, which has been an active field of research for decades. There exist well-established solutions to the problem of enhancing noise-corrupted speech, a wide range of them formulated in the STFT domain, motivated by its mathematical convenience and efficient implementation. The estimation of the clean speech STFT representation has been focused solely on processing the spectral amplitude while leaving the spectral phase untouched. From the perceptual point of view, an accurate estimate of the clean speech spectral amplitude is indispensable. In addition, the spectral phase was found to be perceptually irrelevant by Wang and Lim (1982) (see also Experiment 1.1 in Chapter 1), leading the subsequent research to neglect the information carried by the spectral phase. It is interesting that, at the same time, they state that a reasonable phase estimate could be beneficial if used for refinement of the amplitude estimation. More recent findings (Paliwal *et al.* 2011) revealed contradictory results, conceding perceptual importance to the spectral phase. Following this study, the information carried by the spectral phase gained more and more attention in the application of speech enhancement.

Within the class of STFT-based approaches, there exist various possibilities to formulate the speech enhancement problem. In this chapter, we will consider the class of statistical estimators; more precisely, we will focus on methods that make use of the *minimum mean square error* (MMSE) criterion. Starting from general concepts used in speech enhancement systems (Sections 4.2 and 4.3), we will illustrate how additional knowledge about the spectral phase can be incorporated into the estimation framework in order to push the limits of the phase-insensitive approaches (Section 4.4).

Single Channel Phase-Aware Signal Processing in Speech Communication: Theory and Practice, First Edition.
Pejman Mowlaee, Josef Kulmer, Johannes Stahl, and Florian Mayer.
© 2017 John Wiley & Sons, Ltd. Published 2017 by John Wiley & Sons, Ltd.

4.2 Speech Enhancement in the STFT Domain: General Concepts

A speech enhancement system processes the input speech in order to mitigate the deterioration of quality and intelligibility caused by interfering noise. Figure 4.1 displays the blocks needed to accomplish this task. Under the assumption of additive noise we will use the following signal model in order to formulate the estimation of the clean speech signal:

$$Y(k,l) = R(k,l)e^{j\vartheta(k,l)} = X(k,l) + D(k,l),\tag{4.1}$$

where $Y(k,l)$ is the STFT representation of the noisy observation $y(n)$ at frequency index k and frame l with absolute value $R(k,l)$ and angle $\vartheta(k,l)$. We further define $D(k,l)$ and $X(k,l)$ as the noise and the clean speech time–frequency coefficients with

$$X(k,l) = A(k,l)e^{j\alpha(k,l)}.\tag{4.2}$$

In general, the *speech spectrum estimation* block represents the estimation rules formulated as gain functions that are a function of the *a posteriori* SNR $\zeta(k,l)$ and *a priori* SNR $\xi(k,l)$ defined as

$$\zeta(k,l) = \frac{R^2(k,l)}{\sigma_d^2(k,l)},\tag{4.3}$$

$$\xi(k,l) = \frac{\sigma_x^2(k,l)}{\sigma_d^2(k,l)}.\tag{4.4}$$

To compute the SNRs, the noise *power spectral density* (PSD) $\sigma_d^2(k,l)$ and the speech PSD $\sigma_x^2(k,l)$ also need to be known. Since, in a practical application, they are not directly accessible, the following sections give a short introduction to how these parameters might be obtained. The concrete speech spectral estimators will be discussed afterwards.

4.2.1 *A priori* SNR Estimation

Given a noise PSD estimate, the *a posteriori* SNR is obtained by (4.3). Still, the estimation of the *a priori* SNR proves to be difficult since the speech PSD, given by

$$\sigma_x^2(k,l) = \mathbb{E}(A^2(k,l)),\tag{4.5}$$

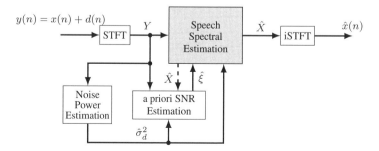

Figure 4.1 The typical blocks needed for STFT speech enhancement: noise PSD estimation, *a priori* SNR estimation, speech spectral estimation.

is not known. Here, $\mathbb{E}(\cdot)$ denotes the expected value operator. Expressing the *a priori* SNR by means of the *a posteriori* SNR yields

$$\xi(k,l) = \mathbb{E}(\zeta(k,l) - 1). \tag{4.6}$$

Formulating the maximum likelihood (ML) estimate of the *a priori* SNR gives us (Ephraim and Malah 1984)

$$\hat{\xi}_{\text{ML}}(k,l) = \frac{1}{M} \sum_{m=0}^{M-1} \max\left[\zeta(k,l-m) - 1, 0\right], \tag{4.7}$$

where M denotes the smoothing window length and the max[·] operator keeps the *a priori* SNR from becoming negative. For $M = 1$ the ML estimator degenerates to the power spectral subtraction method (McAulay and Malpass 1980). The moving average in (4.7) is in practice implemented by a recursive smoothing. Implementations using the *a priori* SNR estimate obtained from (4.7) suffer from the so-called *musical noise*, a phenomenon caused by residual spectral peaks resulting in highly fluctuating tonal components. Since common speech enhancement methods strongly depend on the quality of the *a priori* SNR estimate, its improvement is widely discussed in the literature. In the following, two prominent approaches are presented in order to illustrate the difficulties that arise in *a priori* SNR estimation.

4.2.1.1 Decision-Directed *a priori* SNR Estimation

The decision-directed approach in Ephraim and Malah (1984) combines knowledge about the spectral amplitude and the *a posteriori* SNR by formulating the *a priori* SNR as follows:

$$\xi(k,l) = \frac{\mathbb{E}(A^2(k,l))}{\sigma_d^2(k,l)}. \tag{4.8}$$

The expression in (4.8) is approximated by incorporating the speech spectral coefficient estimates obtained in the previous frame $l - 1$. By introducing a smoothing parameter β_{DD}, combining (4.8) and (4.7) results in

$$\hat{\xi}(k,l) = \beta_{\text{DD}} \frac{\hat{A}^2(k,l-1)}{\sigma_d^2(k,l-1)} + (1 - \beta_{\text{DD}})\max\left[\zeta(k,l) - 1, 0\right], \tag{4.9}$$

where β_{DD} lies in the interval $[0, 1]$ (typically chosen with $\beta_{\text{DD}} > 0.9$). The decision-directed approach helps to suppress musical noise efficiently (Ephraim and Malah 1984). To avoid underestimating the *a priori* SNR, which is in general more critical than its overestimation, the studies in Cappé (1994) and Malah *et al.* (1999) recommend flooring the *a priori* SNR in order to limit the maximum attenuation that is achieved by the spectral estimator.

In order to illustrate the impact of the smoothing parameter on musical noise, Figure 4.2 pictures the log-histograms of the DFT magnitude of the residual noise after applying a Wiener filter with different choices of the smoothing parameter. The histograms ideally follow a Rayleigh distribution, indicating independent real and imaginary parts of the observed DFT coefficients. Heavily pronounced tails of the distributions correspond to spectral outliers that can be associated with musical noise. While small smoothing parameters ($\beta_{\text{DD}} = 0.92$) in the decision-directed approach

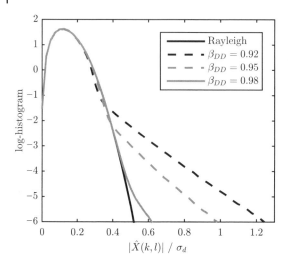

Figure 4.2 Log-histogram in dB plots inspired by Breithaupt *et al.* (2007) for the residual noise DFT magnitude distributions (neglecting DC and Nyquist bin) for different choices of β_{DD} in (4.9).

yield a large number of outliers, higher smoothing parameters contribute to successful attenuation of the musical noise. However, this musical noise reduction for larger smoothing parameters is obtained at the expense of speech distortions. This is why Breithaupt *et al.* (2008a) proposed an approach relying on selective cepstro-temporal smoothing.

4.2.1.2 Cepstro-Temporal Smoothing

From (4.9) we see that the decision-directed approach is sensitive to amplitude onsets, as its outcome depends on the previous frame. To cope with this, Breithaupt *et al.* (2008a) proposed to smooth the *a priori* SNR recursively in the cepstral domain. Since lower cepstral coefficients represent the speech envelope, higher cepstral coefficients (quefrencies) are smoothed more strongly than lower coefficients. It is important to note that higher cepstral coefficients could also represent the fundamental frequency f_0, which is why the smoothing must be applied selectively. Hence, as a first step a fundamental frequency estimate has to be obtained. The amount of recursive smoothing is controlled by a factor similar to β_{DD}. An obvious advantage of this method is that it takes into account the prior knowledge about the underlying speech signal by considering the fundamental frequency for the *a priori* SNR estimation. As a result, less speech distortion is introduced yielding a clearer speech signal. In addition, the *a priori* SNR estimation gets more robust with respect to non-stationary noise types.

4.2.2 Noise PSD Estimation

Until now, we assumed the noise variance to be known. Following its definition,

$$\sigma_d^2(k, l) = \mathbb{E}(|D(k, l)|^2),\qquad(4.10)$$

we know that in a practical application the expected value of $|D(k, l)|^2$ is unknown. Several difficulties arise when trying to solve the noise variance estimation problem:

- Since the noise power may change over time, it is crucial to re-estimate and accordingly update the noise PSD. In order to do so, finite observation intervals are required.

- We only have access to one realization of $D(k,l)$ per frame l and frequency index k. Hence, the expected value is typically approximated by smoothing along time or frequency. This is contradictory to the requirement of an adaptive estimate, yielding a trade-off between the ability to track the evolving characteristics of the noise and a low variance of the noise PSD estimation.
- This procedure is further aggravated by the fact that, in addition to the noise, a speech signal is present in the case of a practical application. Thus, we cannot access the noise signal directly but have to estimate it from the noise-corrupted speech.

A possible solution to this problem is to estimate the noise characteristics at time instances where speech is absent. This concept can only be carried out in combination with reliable voice activity detection. In addition, it is not possible to track the noise PSD during speech presence. In the following, the commonly used noise PSD estimator, based on minimum statistics (Martin 1994), is briefly discussed. This is done in order to show the problems that may arise when trying to obtain the noise PSD.

4.2.2.1 Minimum Statistics

The minimum statistics concept (Martin 1994) relies on the spectro-temporal sparsity of speech, i.e., we assume that noise is dominant between speech harmonics as well as in speech pauses along time. Recursive smoothing of the periodogram of the noisy signal yields an estimate of the signal PSD. By tracking the minima of the smoothed signal power, it serves as the basic estimate of the noise PSD. This estimate is biased, since it tends to underestimate the desired underlying noise PSD. The necessary bias correction was presented in Martin (2001, 2006). In order to cope with changing speech characteristics while still allowing for tracking non-stationary noise, it is also beneficial to apply a signal adaptive smoothing factor. If the noise PSD increases over time, a delay (depending on the smoothing length) is introduced as well as a certain bias towards lower values. State-of-the-art methods that cope with the tracking delay in noise estimation and its bias are presented in, for example, Hendriks *et al.* (2010) and Gerkmann and Hendriks (2012).

4.3 Conventional Speech Enhancement

The core of the speech enhancement system as depicted in Figure 4.1 is the *speech spectral estimation* block. It takes into account the SNR estimates and the statistical model of speech in order to obtain an estimate of the speech spectrum from the noisy observation. As already explained in the introduction of this chapter, the importance of the phase information in this context is a controversial issue. To highlight the peculiarity of the conventional methods, neglecting phase information, Figure 4.3 illustrates a very simplified scheme for such an approach. The noisy phase is used for reconstruction together with an enhanced amplitude, obtained independently from any phase information. Before we depart to phase-sensitive proposals, the statistical model and a phase-insensitive baseline method are explained.

4.3.1 Statistical Model

As a starting point, we will assume that the speech spectral coefficients follow a zero-mean complex Gaussian distribution. This hypothesis is widely used in the

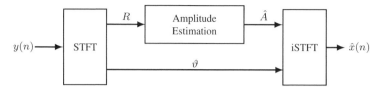

Figure 4.3 Conventional speech enhancement, illustrated by a block diagram. The STFT representation of the noisy signal is given by $Re^{j\vartheta}$, and the estimate of the clean amplitude spectrum A is denoted by $\hat{A}$. The iSTFT($\cdot$) is applied to $\hat{A}e^{j\vartheta}$.

speech enhancement literature, for example Ephraim and Malah (1984), although it is inaccurate, since signals consisting of periodic components are characterized by a non-zero mean (Thomson 1982). Nevertheless, in the following derivations we will assume that the real and imaginary parts of the speech complex spectral coefficients $X(k,l) = X_{\Re}(k,l) + jX_{\Im}(k,l)$ are statistically independent random variables, both following a normal distribution with zero mean. For the sake of readability, we will drop the frame index l and the frequency index k wherever possible in the following. Further, the speech variance σ_x^2 is expected to split equally into real and imaginary parts. The resulting distributions are given by

$$p(X_{\Re}) = \frac{1}{\sqrt{\pi\sigma_x^2}} \exp\left(-\frac{X_{\Re}^2}{\sigma_x^2}\right),$$

$$p(X_{\Im}) = \frac{1}{\sqrt{\pi\sigma_x^2}} \exp\left(-\frac{X_{\Im}^2}{\sigma_x^2}\right). \tag{4.11}$$

In order to keep the notation simple, we forgo the subscripts identifying the *probability density functions* (pdfs) with respect to their corresponding variables, so that their arguments solely fulfill this task. Since the real and imaginary parts are modeled to be statistically independent, their joint distribution is

$$p(X_{\Re}, X_{\Im}) = p(X_{\Re}) \cdot p(X_{\Im}) = \frac{1}{\pi\sigma_x^2} \exp\left(-\frac{X_{\Re}^2 + X_{\Im}^2}{\sigma_x^2}\right). \tag{4.12}$$

By transforming (4.12) from Cartesian to polar coordinates, we obtain the joint distribution of amplitude and phase

$$p(A, \alpha) = \frac{A}{\pi\sigma_x^2} \exp\left(-\frac{A^2}{\sigma_x^2}\right). \tag{4.13}$$

From (4.13), we can get the distributions for the amplitude and the phase by marginalization,

$$p(A) = \int_0^{2\pi} p(A, \alpha)d\alpha = \frac{2A}{\sigma_x^2} \exp\left(-\frac{A^2}{\sigma_x^2}\right), \tag{4.14}$$

$$p(\alpha) = \int_0^{\infty} p(A, \alpha)dA = \frac{1}{2\pi}, \tag{4.15}$$

where (4.14) is known as the Rayleigh distribution and (4.15) denotes the uniform distribution of the phase.

In the case of additive noise the probability density function of the noisy observation Y, conditioned on the clean speech's amplitude A, is of interest. Assuming independent, Gaussian-distributed real and imaginary parts of the noise spectral coefficients, the resulting distribution of Y conditioned on A and α is

$$p(Y|A, \alpha) = \frac{1}{\pi \sigma_d^2} \exp\left(-\frac{|Y - Ae^{j\alpha}|^2}{\sigma_d^2}\right), \tag{4.16}$$

which results in a *Rician* distribution for the conditional magnitude R,

$$p(R|A) = \frac{2R}{\sigma_d^2} \exp\left(-\frac{R^2 + A^2}{\sigma_d^2}\right) I_0\left(\frac{2AR}{\sigma_d^2}\right), \tag{4.17}$$

with $I_v(\cdot)$ denoting the modified Bessel function of the first kind and order v, derived from the integral (Gradshteyn and Ryzhik 2007)

$$I_v(z) = \frac{1}{2\pi} \int_0^{2\pi} \cos(\alpha v) \exp(z \cos(\alpha)) \, d\alpha. \tag{4.18}$$

It has been shown that the Rayleigh distribution does not necessarily fit the speech amplitude's distribution well (Lotter and Vary 2004). The Rayleigh prior results from Gaussian real and imaginary parts. Other assumptions on the components, such as Gamma or Laplace priors yield different results (Martin 2005) regarding the speech enhancement performance. Several alternatives have been proposed; in particular, the proposal in Breithaupt *et al.*, (2008b) is of interest in this chapter:

$$p(A) = \frac{2}{\Gamma(\mu)} \left(\frac{\mu}{\sigma_x^2}\right)^\mu A^{2\mu-1} \exp\left(-\frac{\mu}{\sigma_x^2} A^2\right). \tag{4.19}$$

The shape parameter μ allows for modeling numerous prior distributions including the Rayleigh for the case $\mu = 1$ as well as super-Gaussian priors, achieved with values $\mu < 1$.

4.3.2 Short-Time Spectral Amplitude Estimation

Motivated by the perceptual importance of the *short-time spectral amplitude* (STSA), Ephraim and Malah (1984) proposed to estimate the STSA in the MMSE sense, yielding a commonly used estimation rule that can be considered as the baseline method in this chapter, since the approaches described later will extend its basic idea in a phase-aware sense.

Following the assumption that the noisy time domain observation $y(n)$ and its spectral representation Y bear the same information, as well as that the N_{DFT} spectral components form a statistically independent set of observations, the problem of estimating the speech spectral amplitude A from Y is justified. The cost function to minimize is the Bayesian mean square error $\text{BMSE}(\cdot)$ given by

$$\text{BMSE}(\hat{A}) = \mathbb{E}((A - \hat{A})^2), \tag{4.20}$$

yielding the MMSE estimator (Kay 1993)

$$\hat{A} = \int_0^\infty \int_0^{2\pi} A p(A, \alpha | Y) \, d\alpha dA$$

$$= \int_0^\infty A p(A | Y) \, dA$$

$$= \mathbb{E}(A | Y), \tag{4.21}$$

which can be rewritten by using Bayes' theorem $p(B|C) = \frac{p(C|B)p(B)}{p(C)}$:

$$\hat{A} = \frac{\int_0^\infty \int_0^{2\pi} A p(Y|A, \alpha) p(A, \alpha) \, d\alpha dA}{\int_0^\infty \int_0^{2\pi} p(Y|A, \alpha) p(A, \alpha) \, d\alpha dA}. \tag{4.22}$$

We obtain the desired MMSE estimator of the STSA by inserting (4.13) and (4.16) into (4.22),

$$\hat{A} = \Gamma(1.5) \frac{\sqrt{\nu}}{\zeta} \exp\left(-\frac{\nu}{2}\right) \left((1+\nu) I_0 \left(\frac{\nu}{2}\right) + \nu I_1 \left(\frac{\nu}{2}\right) \right) R, \tag{4.23}$$

with

$$\nu = \frac{\xi}{1+\xi} \zeta, \tag{4.24}$$

and $\Gamma(\cdot)$ denoting the Gamma function.

Since low magnitudes are crucial for intelligibility, often compressed functions of the magnitude A are estimated (e.g. logarithmically compressed, Ephraim and Malah 1985). This helps to put emphasis on errors in the estimation of small amplitudes rather than in the estimation of larger amplitudes. This is why it can be perceptually advantageous to estimate A^β instead of A (You *et al.* 2005). The compression function $c(\cdot)$ can in general be expressed by (Breithaupt *et al.* 2008b)

$$c(x) = x^\beta, \tag{4.25}$$

where β denotes the compression parameter. A very flexible version of the MMSE-STSA estimator can be obtained by utilizing the parametric distribution given in (4.19) together with the compression function in (4.25), yielding (Breithaupt *et al.* 2008b)

$$\hat{A} = c^{-1}(c(A))$$

$$= \sqrt{\frac{1}{\zeta} \frac{\xi}{\mu + \xi}} \underbrace{\left(\frac{\Gamma\left(\mu + \frac{\beta}{2}\right)}{\Gamma(\mu)} \frac{{}_1F_1\left(1 - \mu - \frac{\beta}{2}, 1; -\nu\right)}{{}_1F_1(1 - \mu, 1; -\nu)} \right)^{\frac{1}{\beta}}}_{G_{\text{(Breithaupt } et\ al.\ 2008b)}} R, \tag{4.26}$$

where in this case $\nu = \frac{\xi}{\mu + \xi} \zeta$ and ${}_1F_1(\cdot, \cdot; \cdot)$ is the confluent hypergeometric function. Provided the phase is uniformly distributed, its MMSE estimate was shown to be the noisy phase (Ephraim and Malah 1984). Therefore, the estimate of the complex speech DFT coefficients is given by

$$\hat{X}_{\text{(Breithaupt } et\ al.\ 2008b)} = G_{\text{(Breithaupt } et\ al.\ 2008b)} Y = G_{\text{(Breithaupt } et\ al.\ 2008b)} R e^{j\vartheta}, \tag{4.27}$$

Table 4.1 Spectral amplitude estimators that are special cases of the parametrized estimator in (4.26).

Parameters	Estimator
$\mu = 1, \beta = 1$	Ephraim and Malah (1984), (4.23)
$\mu = 1, \beta \rightarrow 1$	Ephraim and Malah (1985)
$\mu = 1, \beta > 0$	You *et al.* (2005)
$\mu > 0, \beta = 1$	Andrianakis and White (2009); Erkelens *et al.* (2007)

where G is the so-called gain function, more precisely defined by its subscript. The parametrization in (4.26) unifies several MMSE-STSA estimators derived from different assumptions. Table 4.1 lists the extreme cases achieved by the parameterized MMSE-STSA estimator.

4.4   Phase-Sensitive Speech Enhancement

So far, independence of the complex coefficient's real and imaginary parts was assumed. This assumption yields a uniformly distributed phase prior, as shown in (4.15). However, as already argued in Chapters 2 and 3, the spectral phase follows a structure that can be described by a non-uniform distribution. By deciding to take advantage of phase estimates for better signal estimation, one has basically three options to incorporate this additional source of information:

1) Estimate the clean speech's spectral phase and apply it at the synthesis stage, replacing the noisy phase.
2) Use a phase estimate as additional information for the amplitude estimation.
3) Jointly estimate amplitude and phase in order to profit from the contrasting sources of information.

The first scenario is covered in Chapter 3. In this chapter we will focus on the remaining possibilities to integrate knowledge about the spectral phase. Still, we will need the estimators described in Chapter 3 in order to gather information about the unknown phase. Based on this, we can incorporate the phase into the overall estimation in a second step. Figure 4.4 illustrates the differences in the three approaches in block diagrams.

4.4.1   Phase Estimation for Signal Reconstruction

The phase estimation methods demonstrated in Chapter 3 have been successfully incorporated into speech enhancement systems at the signal reconstruction stage. One major issue in phase-insensitive speech enhancement is that, in general, only perceived speech quality can be increased, while intelligibility is degraded (Loizou and Kim 2011). Combinations of phase estimators and phase-unaware amplitude estimators show that the negative impact of the amplitude estimators on intelligibility can be partially compensated by using additional phase enhancement (Mowlaee and Kulmer 2015).

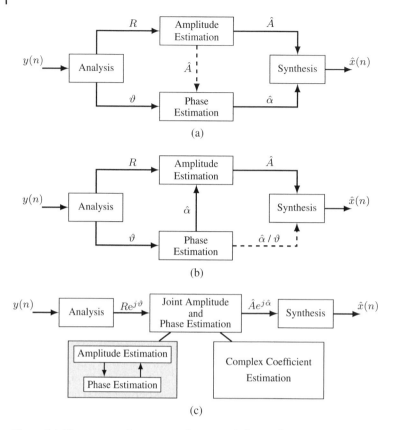

Figure 4.4 Three ways to incorporate the spectral phase information into the overall spectral estimation procedure. (a) Use independently obtained estimates for *A* and *α* for reconstruction. Some of the phase estimators depicted in Chapter 3 need a spectral amplitude estimate, which is in general not derived from the same cost function and does not comprise any phase information in this scenario. (b) Phase information is used in order to refine the amplitude estimate. It is optional to use the phase estimate or the noisy phase for reconstruction (indicated by the dashed line, see Section 4.4.2). (c) Amplitude and phase are obtained jointly; there are several ways to accomplish this, and both estimates are employed for synthesis.

4.4.2 Spectral Amplitude Estimation Given the STFT Phase

Given a phase estimate, a first step to incorporating the additional knowledge into the overall speech spectral estimation is to reformulate the MMSE criterion in (4.21) by assuming the phase to be known:

$$\hat{A} = \mathbb{E}(A|Y, \alpha) = \int_0^\infty Ap(A|Y, \alpha)\, dA. \tag{4.28}$$

Similar to Section 4.3.2, by applying Bayes' theorem we obtain the following integral to solve:

$$\hat{A} = \frac{\int_0^\infty Ap(Y|A, \alpha)p(A, \alpha)\, dA}{\int_0^\infty p(Y|A, \alpha)p(A, \alpha)\, dA}, \tag{4.29}$$

whereas $p(\alpha)$ can now be factored out, which (with the same statistical model as before) results in

$$\hat{A} = \frac{\int_0^\infty A p(Y|A,\alpha) p(A)\, dA}{\int_0^\infty p(Y|A,\alpha) p(A)\, dA}, \tag{4.30}$$

$$= \frac{\int_0^\infty A^2 \exp\left(-A^2\left(\frac{1}{\sigma_d^2} + \frac{1}{\sigma_x^2}\right) + \frac{2AR}{\sigma_d^2}\cos(\vartheta - \alpha)\right)\, dA}{\int_0^\infty A \exp\left(-A^2\left(\frac{1}{\sigma_d^2} + \frac{1}{\sigma_x^2}\right) + \frac{2AR}{\sigma_d^2}\cos(\vartheta - \alpha)\right)\, dA}; \tag{4.31}$$

this can be solved by using equation (3.462) in Gradshteyn and Ryzhik (2007),

$$\int_0^\infty x^{\nu-1} \exp(-\beta x^2 - \gamma x)\, dx = (2\beta)^{-\frac{\nu}{2}} \Gamma(\nu) \exp\left(\frac{\gamma^2}{8\beta}\right) D_{-\nu}\left(\frac{\gamma}{\sqrt{2\beta}}\right), \tag{4.32}$$

with $D_{-\nu}(\cdot)$ and $\Gamma(\cdot)$ denoting the parabolic cylinder function and the gamma function respectively. The resulting estimator of the spectral amplitude given the clean spectral phase is therefore

$$\hat{A} = \sqrt{\frac{2}{\zeta}\frac{\xi}{\xi+1}} \frac{D_{-3}\left(-\sqrt{2\zeta\frac{\xi}{\xi+1}}\cos(\Delta\phi)\right)}{D_{-2}\left(-\sqrt{2\zeta\frac{\xi}{\xi+1}}\cos(\Delta\phi)\right)} R, \tag{4.33}$$

where $\Delta\phi = \vartheta - \alpha$.

Again, a more general (i.e. parameterized amplitude distribution and compression function) version of the MMSE-STSA estimator for given phase can be found in the literature (Gerkmann and Krawczyk 2013):

$$\hat{A} = \left(\mathbb{E}(A^\beta | R, \vartheta, \alpha)\right)^{\frac{1}{\beta}} = \left(\int_0^\infty A^\beta p(A|R,\vartheta,\alpha)\, dA\right)^{\frac{1}{\beta}}$$

$$= \underbrace{\sqrt{\frac{2}{\zeta}\frac{\xi}{\mu+\xi}}\left[\frac{\Gamma(2\mu+\beta)}{\Gamma(2\mu)}\frac{D_{-(2\mu+\beta)}\left(-\sqrt{2\zeta\frac{\xi}{\mu+\xi}}\cos(\Delta\phi)\right)}{D_{-(2\mu)}\left(-\sqrt{2\zeta\frac{\xi}{\mu+\xi}}\cos(\Delta\phi)\right)}\right]^{\frac{1}{\beta}}}_{G_{(\text{Gerkmann and Krawczyk 2013})}} R. \tag{4.34}$$

For $\beta = 1$ and $\mu = 1$, (4.34) reduces to (4.33). Both have in common that they are a function of the cosine of the phase deviation $\cos(\Delta\phi)$. In a real scenario the clean phase α is unknown. This is why in practice a phase estimate $\hat{\alpha}$ has to be utilized. Gerkmann and Krawczyk (2013) proposed using STFTPI Krawczyk and Gerkmann (2012) for the estimation of α. This combination turned out to be especially advantageous in the case where the phase estimate is only used for amplitude estimation while for reconstruction the noisy phase is employed. The final complex coefficients are therefore computed by using

$$\hat{X}_{(\text{Gerkmann and Krawczyk 2013})} = G_{(\text{Gerkmann and Krawczyk 2013})} R e^{j\vartheta}. \tag{4.35}$$

Ephraim and Malah (1983) proposed to solve the problem of obtaining $\cos(\Delta\phi)$ by deriving an MMSE estimate of $\cos(\Delta\phi)$ directly, bypassing the phase estimation itself. Under the assumption of given amplitude and uniform phase prior $p(\alpha) = \frac{1}{2\pi}$ with $\alpha \in [-\pi, \pi[$, we get

$$\widehat{\cos(\Delta\phi)} = \mathbb{E}(\cos(\Delta\phi)) = \frac{\int_0^{2\pi} \cos(\Delta\phi) p\,(Y|A, \alpha) p\,(\alpha)\,d\alpha}{\int_0^{2\pi} p\,(Y|A, \alpha) p\,(\alpha)\,d\alpha}$$

$$= \frac{I_1\left(2\frac{RA}{\sigma_d^2}\right)}{I_0\left(2\frac{RA}{\sigma_d^2}\right)}. \tag{4.36}$$

The highly non-linear equations given in (4.33) and (4.36) can be solved numerically only, and so far no closed-form solution has been published. Hence, in Ephraim and Malah (1983) the SNR dependence in (4.36) is used in order to apply asymptotic approximations of $\widehat{\cos(\Delta\phi)}$ for the amplitude estimation. The reconstruction, however, in this case also needs to be carried out by employing the noisy phase since $\cos^{-1}(\cos(\Delta\phi))$ yields ambiguous results for $\Delta\phi$.

Figure 4.5 illustrates the impact of the choice for α in (4.33) by means of the spectrograms of the enhanced speech. To this end we chose a TIMIT (Garofolo *et al.* 1993) utterance from the train set, corrupted with white noise at 0 dB. In the case where the clean phase is accessible for amplitude estimation we observe several benefits when compared to the utilization of the noisy phase:

- Speech onsets are restored (i.e., transients, highlighted by the gray rectangles in Figure 4.5).
- Speech harmonics are reconstructed while the information between is preserved (black, solid rectangles).
- Artifacts introduced due to wrong harmonic model-order classification are not showing up (see Figure 4.5(d), dashed rectangle).

The phase estimation approach in (e) leads to fewer harmonic artifacts than in (d) (at the expense of less reconstructed harmonic structure), with slightly improved noise suppression compared to the phase-unaware baseline method in (c).

4.4.3 Iterative Closed-Loop Phase-Aware Single-Channel Speech Enhancement

From (4.33) and (4.34) there is the potential to incorporate knowledge about the clean speech phase into the overall speech spectral estimation by the term $\cos(\Delta\phi)$. Mowlaee and Saeidi (2013) presented an iterative approach that combines an amplitude estimator similar to (4.34) together with the geometric phase estimator in Mowlaee *et al.* (2012).

To provide an initial amplitude estimate, the conventional Wiener filter is applied. This estimate is exploited to estimate a phase spectrum using the geometry method explained in Chapter 3 for speech spectral components with a local SNR lower than 6 dB (following the observations made in Vary 1985). To refine the spectral amplitude estimate, the enhanced phase spectrum is fed back into an amplitude estimator, now phase-aware as in (4.34). From Chapter 2 we know that the spectro-temporal dependencies that arise from the redundancy of the STFT analysis (finite frame lengths, overlapping frames) are

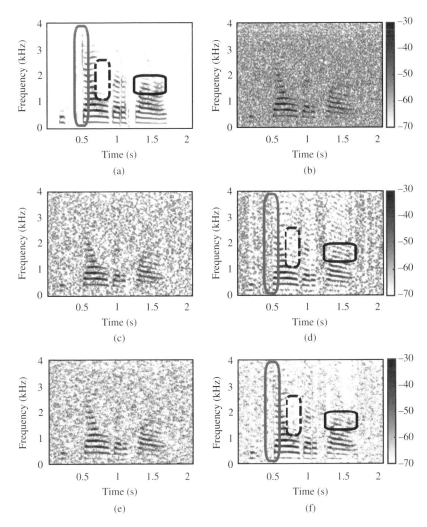

Figure 4.5 Spectrograms of the clean speech (a), noise-corrupted speech (b), and enhanced speech signals (c). $\alpha = \vartheta$, leading (4.33) to reduce to (4.23). (d) Phase estimate obtained by Krawczyk and Gerkmann (2014), (e) phase estimate obtained by Kulmer and Mowlaee (2015), and (f) (4.33) with clean phase given. Depending on the choice for α, harmonics are restored and artifacts introduced.

degraded in the case of amplitude-only enhancement. This induced *inconsistency* may be compensated by employing information about the clean phase into the reconstruction and/or the amplitude estimation process. As a consequence, for the iterative approach a consistency constraint can be formulated as a stopping criterion (Griffin and Lim 1984),

$$I(\hat{\mathbf{X}}_{\hat{\alpha}^{(i)}}^{(i)}) = \hat{\mathbf{X}}_{\hat{\alpha}^{(i)}}^{(i)} - \mathcal{G}(\hat{\mathbf{X}}_{\hat{\alpha}^{(i)}}^{(i)}) \qquad i: \text{iteration index,} \qquad (4.37)$$

where $\mathcal{G} = \text{STFT}(\text{iSTFT}(\cdot))$ operator. Since we discarded the frame index l and frequency index k for the sake of notational simplicity, in this section bold letters denote a

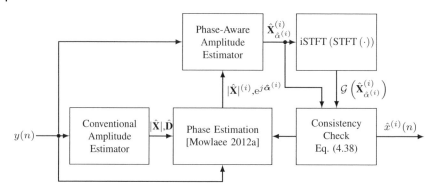

Figure 4.6 The iterative closed loop method (Mowlaee and Saeidi 2013). The estimate $\hat{\mathbf{X}}_{\hat{\alpha}^{(i)}}^{(i)}$ is given by (4.34) and the estimated phase is obtained by the geometry method presented in Mowlaee *et al.* (2012).

full complex STFT. The subscript $\hat{\alpha}^{(i)}$ indicates that the spectral amplitude estimate is achieved by a phase-sensitive estimator. The stopping criterion is formulated as follows:

$$\epsilon^{(i)} = \frac{\left|\Delta_{\mathcal{I}}^{(i)}\right|}{\left\|\mathcal{I}\left(\hat{\mathbf{X}}_{\hat{\alpha}^{(i)}}^{(i)}\right)\right\|_2^2} < \epsilon_{\text{th}}, \tag{4.38}$$

where the threshold ϵ_{th} is set to $\epsilon_{\text{th}} = 0.05$ and $\Delta_{\mathcal{I}}^{(i)}$ is the change in inconsistency from one iteration to the next,

$$\Delta_{\mathcal{I}}^{(i)} = \left\|\mathcal{I}\left(\hat{\mathbf{X}}_{\hat{\alpha}^{(i)}}^{(i)}\right)\right\|_2^2 - \left\|\mathcal{I}\left(\hat{\mathbf{X}}_{\hat{\alpha}^{(i-1)}}^{(i-1)}\right)\right\|_2^2. \tag{4.39}$$

The final enhanced speech is therefore given by

$$\hat{x}(n) = \hat{x}^{(I)}(n) = \text{iSTFT}\left(\hat{\mathbf{X}}_{\hat{\alpha}^{(I)}}^{(I)}\right), \tag{4.40}$$

where I is the first iteration that fulfills $\epsilon^{(I)} < \epsilon_{\text{th}}$.

In (4.37), $\mathcal{G}(\hat{\mathbf{X}}_{\hat{\alpha}^{(i)}}^{(i)})$ always represents a *consistent* STFT while $\hat{\mathbf{X}}_{\hat{\alpha}^{(i)}}^{(i)}$ is the modified STFT, which is not consistent *per se*. Across iterations, $\mathcal{I}(\hat{\mathbf{X}}_{\hat{\alpha}^{(i)}}^{(i)})$ decreases and gets saturated, indicating restored spectro-temporal correlation. Figure 4.6 illustrates the scheme of the method and Figure 4.7 shows the spectrograms of the enhanced speech together with the normalized change in inconsistency across iterations.

4.4.4 Incorporating Voiced/Unvoiced Uncertainty

The proposed phase reconstruction method in Krawczyk and Gerkmann (2012) relies on a harmonic model of speech. This is mostly valid for voiced speech. Thus, it seems convenient to incorporate the obtained phase estimate only for voiced regions, while a phase-insensitive amplitude estimator is used for unvoiced speech segments. This is achieved by setting the phase deviation $\Delta\phi$ to zero, leading the estimator in (4.34)

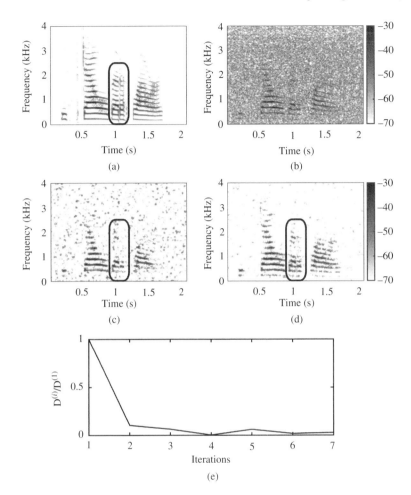

Figure 4.7 Spectrograms of the iterative method and relative inconsistency across iterations. (a) Clean speech, (b) noise corrupted speech, (c) iterative blind, (d) iterative method with noise magnitude assumed to be known for the initial geometry-based phase estimate, (e) the normalized change in inconsistency across iterations. The rectangle shows the speech reconstruction provided a reliable phase estimate is given.

to boil down to (4.26). To weight the phase-aware and the phase-unaware amplitude estimates (denoted by the subscripts), the voicing probability P_{H_v} is employed (Krawczyk *et al.* 2013):

$$\hat{A}_{(\text{Krawczyk } \textit{et al.} \ 2013)} = P_{H_v} |\hat{X}_{(\text{Gerkmann and Krawczyk 2013})}|$$
$$+ (1 - P_{H_v}) |\hat{X}_{(\text{Breithaupt } \textit{et al.} \ 2008b)}|, \tag{4.41}$$

with $0 \le P_{H_v} \le 1$. This weighing helps to avoid employing an unreliable phase estimate at unvoiced frames, while profiting from the additional phase information at voiced frames.

4.4.5 Uncertainty in Prior Phase Information

The artifacts introduced by incorporating a phase estimate strictly relying on the sinusoidal model of speech have aggravated the choice of a reconstruction phase so far. The deterministic model does not fit unvoiced sounds, and certainly not speech absence. The solution in (4.41) and the SNR dependency of the chosen estimation rule in Section 4.4.3 are the first attempts to cope with this inaccuracy. By explicitly imposing uncertainty on the provided phase estimate it is possible to further generalize the estimation of the complex speech spectral coefficients. Representing these considerations in terms of MMSE estimation (amplitude compressed), we obtain (Gerkmann 2014b):

$$\hat{X}^{(\beta)} = \mathbb{E}(A^\beta \exp(j\alpha)|Y, \alpha_\mu) = \int_0^\infty \int_0^{2\pi} A^\beta \exp(j\alpha) p\left(A, \alpha | Y, \alpha_\mu\right) d\alpha dA, \qquad (4.42)$$

where α_μ indicates the uncertain prior phase information. In the context of the Bayesian interpretation of the phase estimate it is important to note that given the clean phase, the phase estimate bears no information, so that

$$p\left(Y|A, \alpha, \alpha_\mu\right) = p\left(Y|A, \alpha\right). \qquad (4.43)$$

The joint probability of amplitude and phase (provided they are statistically independent) illustrates the impact of the phase estimate:

$$p\left(A, \alpha, \alpha_\mu\right) = p(A)p(\alpha, \alpha_\mu) = p(A)p(\alpha_\mu)p(\alpha|\alpha_\mu). \qquad (4.44)$$

The conditional term $p(\alpha|\alpha_\mu)$ captures the uncertainty of the phase estimate. Given this expression, Gerkmann (2014b) suggests utilizing a von Mises distribution to model the distribution of the clean phase around its initial estimate:

$$p(\alpha|\alpha_\mu) = \frac{\exp(\kappa \cos(\alpha - \alpha_\mu))}{2\pi I_0(\kappa)}. \qquad (4.45)$$

By utilizing the parameterized prior distribution of the spectral amplitude in (4.19), the solution to (4.42) is given by

$$\hat{X}^{(\beta)} = \left(\sqrt{\frac{1}{2}\frac{\xi}{\mu + \xi}\sigma_d^2}\right)^\beta \frac{\Gamma(2\mu + \beta)}{\Gamma(2\mu)} \frac{\int_0^{2\pi} \exp\left(j\alpha + \frac{\nu^2}{4}\right) D_{(-2\mu-\beta)}(\nu)p(\alpha|\alpha_\mu)d\alpha}{\int_0^{2\pi} \exp\left(\frac{\nu^2}{4}\right) D_{(-2\mu)}(\nu)p(\alpha|\alpha_\mu)d\alpha}, \qquad (4.46)$$

with

$$\nu = -\sqrt{2\zeta\frac{\xi}{\xi + \mu}} \cos\left(\Delta\phi\right). \qquad (4.47)$$

In order to compensate for the amplitude compression β, the final speech spectral coefficient estimates are obtained by

$$\hat{X} = \frac{\left|\hat{X}^{(\beta)}\right|^{\frac{1}{\beta}}}{\left|\hat{X}^{(\beta)}\right|}\hat{X}^{(\beta)}. \qquad (4.48)$$

Instead of solving the integrals analytically in Gerkmann (2014b), it is proposed to numerically solve the integrals in (4.46).

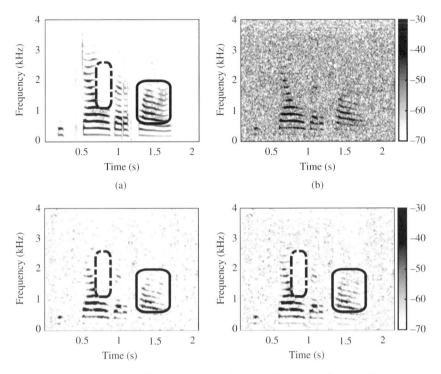

Figure 4.8 (a) Clean speech, (b) noise corrupted speech, (c) estimated phase, (d) oracle phase.

The concentration parameter κ controls the impact of the phase estimate α_μ on the estimation result. By setting it to zero the estimator in (4.26) is achieved. The voicing probability P_{H_v} can be interpreted as a reliability measure of a phase estimate that originates in the sinusoidal model of speech. Therefore, the literature (e.g., Gerkmann 2014a,b; Vanambathina and Kumar 2016) suggests assigning κ according to P_{H_v}:

$$\kappa(k,l) = \begin{cases} 4P_{H_v}(l) & \text{if } \frac{kf_s}{N_{\text{DFT}}} < 4000 \text{ Hz}, \\ 2P_{H_v}(l) & \text{if } \frac{kf_s}{N_{\text{DFT}}} \geq 4000 \text{ Hz}. \end{cases} \tag{4.49}$$

Smaller values of κ at higher frequencies impose a less pronounced harmonic structure within this frequency range. The specific choice of this assignment is chosen to optimize the instrumentally predicted perceived speech quality (see Chapter 6 for more information). Figure 4.8 gives an insight into the potential of the *Bayesian estimator of complex spectral coefficients given uncertain prior phase information* (CUP) (Gerkmann 2014b) estimator and its treatment of the provided phase estimate.

4.4.6 Stochastic–Deterministic MMSE-STFT Speech Enhancement

So far, the methods mentioned have made only little use of known speech characteristics like its composition of periodic (i.e. deterministic if we assume the model parameters to be known) and stochastic components. An effort in this direction was made by incorporating the knowledge about the spectral phase implied by the harmonic model of speech, such as in Sections 4.4.5 and 4.4.2. The hypothesis of deterministic

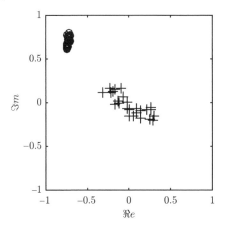

Figure 4.9 Spectral coefficients of a purely stochastic signal (data points centered around zero) and a deterministic signal with uncertainty (decentered data assembly). The phase of the deterministic signal is normalized (achieved by linear phase removal, as in Chapter 3). Figure inspired by Hendriks *et al.* (2007).

components in the speech signal entails that the assumption of a zero-mean complex Gaussian distribution of the speech DFT coefficients is no longer valid (Thomson 1982). Instead, this enforces a complex mean value for the resulting probability density function of the spectral coefficients. Departing from this insight, in Hendriks *et al.* (2007), two different estimates (relying either on the stochastic or on the deterministic model of speech) are combined (see Figure 4.9[1]). This is achieved by incorporating the probabilities of the following three scenarios:

1) presence of a deterministic speech signal
2) presence of a stochastic speech signal
3) absence of speech

in the manner of MMSE estimation under signal-presence uncertainty as presented in Ephraim and Malah (1984).

Based on the stochastic–deterministic (SD) nature of the speech signal we get (McCallum and Guillemin 2013)

$$x_{SD}(n) = \underbrace{\sum_{h=1}^{H} A_h \cos\left(2\pi h \frac{f_0}{f_s} n + \phi_h\right)}_{x_D(n)} + u(n), \tag{4.50}$$

where the stochastic zero-mean term $u(n)$ represents unvoiced speech sounds, hence expanding the purely deterministic model of a harmonic signal ($x_D(n)$) as presented in Chapter 3. Exploiting the linearity of the DFT, the resulting STFT is the sum of windowed deterministic components ($X_{D,W}$) and a (by definition) zero-mean complex Gaussian distributed variable U:

$$X = X_{D,W} + U. \tag{4.51}$$

The adapted joint probability density function of the spectral amplitude and phase has the form

$$p(A, \alpha) = \frac{A}{\pi \sigma_{x,SD}^2} \exp\left(-\frac{|Ae^{j\alpha} - A_\mu e^{j\alpha_\mu}|^2}{\sigma_{x,SD}^2}\right), \tag{4.52}$$

---

1 The MATLAB® implementation for this experiment can be found in *Fig4_9.m* provided in the PhaseLab Toolbox (see Appendix for more details).

where A_μ and α_μ denote the deterministic parts of the speech (i.e. the complex mean value $X_{D,W} = A_\mu e^{j\alpha_\mu}$). Here, the uncertainty measure $\sigma^2_{x,\text{SD}}$ is defined as

$$\sigma^2_{x,\text{SD}} = \mathbb{E}(|X - A_\mu e^{j\alpha_\mu}|^2) = 2\mathbb{E}(|U|^2). \tag{4.53}$$

Again, by solving for $\mathbb{E}(A|Y)$ we get a similar expression to (4.23):

$$\hat{A}_{\text{(McCallum and Guillemin 2013)}} =$$

$$\underbrace{\Gamma(1.5)\sqrt{\frac{\xi_{\text{SD}}}{(\xi_{\text{SD}}+1)\zeta}} \exp\left(-\frac{\nu_{\text{SD}}}{2}\right)\left((1+\nu_{\text{SD}})I_0\left(\frac{\nu_{\text{SD}}}{2}\right)+\nu_{\text{SD}}I_1\left(\frac{\nu_{\text{SD}}}{2}\right)\right)R}_{G_{\text{(McCallum and Guillemin 2013)}}}, \tag{4.54}$$

with

$$\nu_{\text{SD}} = \frac{1}{1+\xi_{\text{SD}}}\left(\xi_{\text{SD}}\zeta + \eta + 2\sqrt{\eta\zeta}\cos(\alpha_\mu - \vartheta)\right). \tag{4.55}$$

In this context the subscript of ξ_{SD} clarifies that it now differs fundamentally from the *a priori* SNR defined in (4.8), since

$$\xi_{\text{SD}} = \frac{\sigma^2_{x,\text{SD}}}{\sigma^2_d}, \qquad \eta = \frac{A^2_\mu}{\sigma^2_{x,\text{SD}}}. \tag{4.56}$$

In (4.56), ξ_{SD} and η can be interpreted as the *a priori signal uncertainty to noise ratio* and the *signal prediction to uncertainty ratio*, respectively (McCallum and Guillemin 2013). It can be seen that for large values of η the main contribution to the power of the speech signal comes from the deterministic A^2_μ. This means that $\sigma^2_{x,\text{SD}}$ expresses the uncertainty of A^2_μ. Considering the posterior probability

$$p(A,\alpha|R,\vartheta) = \frac{p(R,\vartheta|A,\alpha)p(A,\alpha)}{\int_0^\infty \int_0^{2\pi} p(R,\vartheta|A,\alpha)p(A,\alpha)\,d\alpha dA}$$

$$= \frac{A\exp\left(-\frac{|A\exp(j\alpha)-\tilde{Y}|^2}{\sigma^2_{x,\text{SD}}\sigma^2_d/(\sigma^2_{x,\text{SD}}+\sigma^2_d)}\right)}{\int_0^\infty A\int_0^{2\pi}\exp\left(-\frac{|A\exp(j\alpha)-\tilde{Y}|^2}{\sigma^2_{x,\text{SD}}\sigma^2_d/(\sigma^2_{x,\text{SD}}+\sigma^2_d)}\right)d\alpha dA}, \tag{4.57}$$

the problem is the estimation of the expected value of the magnitude of a non-zero-mean complex Gaussian random variable with *a posteriori* mean (McCallum and Guillemin 2013)

$$\tilde{Y} = \frac{\xi_{\text{SD}}}{\xi_{\text{SD}}+1}Re^{j\vartheta} + A_\mu e^{j\alpha_\mu}\left(1 - \frac{\xi_{\text{SD}}}{\xi_{\text{SD}}+1}\right). \tag{4.58}$$

From (4.58) and (4.54) it can be seen that for $A_\mu = 0$ the solution reduces to the solution in (4.23). In the context of phase-aware speech enhancement, (4.58) is especially of interest in the case $A_\mu \neq 0$. Here, the weight of each source of information (i.e. the observation and the prior knowledge about speech) is adjusted by the *a priori* signal uncertainty ξ_{SD}. In this case, not only the magnitude of $\tilde{Y}$ but also its phase is changed. If we assume that $A = \hat{A}$, the maximum likelihood estimator of the phase α will differ from the noisy phase, which will be shown in the following.

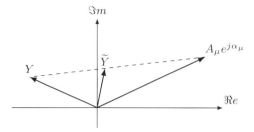

Figure 4.10 Graphical representation of (4.58), illustrating that, depending on ξ_{SD}, the phase and the amplitude of $\tilde{Y}$ are altered. The resulting phasor is a weighted sum of the complex mean $A_\mu e^{j\alpha_\mu}$ and the noisy observation Y. Therefore, the ML estimate of the spectral phase is not the noisy phase any more but a value between the noisy phase ϑ and the prior information $\hat{\alpha}_\mu$, depending on the certainty of the deterministic model (McCallum and Guillemin 2013).

Maximizing the likelihood function with respect to α is expressed by

$$\hat{\alpha} = \arg\max_{\alpha} \log\left(p\left(A, \alpha | R, \vartheta\right)\right), \tag{4.59}$$

which can be solved by setting the derivative of (4.57) with respect to α to zero,

$$\frac{\partial \log\left(p\left(A, \alpha | R, \vartheta\right)\right)}{\partial \alpha} = 0, \tag{4.60}$$

where only the terms depending on α need to be considered, so that the maximum likelihood estimate of α is one of the solutions to

$$\sin\left(\hat{\alpha} - \angle\tilde{Y}\right) = 0. \tag{4.61}$$

By evaluating the second derivative of (4.57), the maximum likelihood estimate of the phase is finally given by

$$\hat{\alpha}_{(\text{McCallum and Guillemin 2013})} = \angle\tilde{Y}. \tag{4.62}$$

The overall estimation can therefore be expressed as

$$\hat{X}_{(\text{McCallum and Guillemin 2013})} = G_{(\text{McCallum and Guillemin 2013})} R e^{j\hat{\alpha}_{(\text{McCallum and Guillemin 2013})}}. \tag{4.63}$$

Two extreme cases of (4.62) are of special interest:

1) $\frac{\xi_{SD}}{\xi_{SD}+1} = 1$ and $A_\mu = 0$, indicating high randomness of the prior knowledge;
2) $\frac{\xi_{SD}}{\xi_{SD}+1} = 0$ and $A_\mu \neq 0$, meaning that the phase estimate is defined only by the prior knowledge.

The first case is consistent with Section 4.3.2, whereas the second case illustrates the ability of the formulated estimator to incorporate prior knowledge about the underlying speech signal. Figure 4.10 illustrates the weighting between the deterministic and stochastic observations in (4.58).

4.4.6.1 Obtaining the Speech Parameters

In order to benefit from prior knowledge about the speech signal it is crucial to obtain reliable parameters of the model in (4.50). The possibilities to achieve f_0, H, ϕ_h, A_h are manifold (e.g., Kay 1993; Thomson 1982) and therefore outside the scope of this chapter. Given reliable estimates, it is therefore of special interest how to integrate them into an STFT framework in the form of A_μ, α_μ, and $\sigma_{x,\,SD}$. Since we have

$$\sigma_{x,\,SD}^2 = \mathbb{E}(|Ae^{j\alpha} - A_\mu e^{j\alpha_\mu}|^2), \tag{4.64}$$

McCallum and Guillemin (2013) proposed transforming the deterministic portion of the model in (4.50) into the STFT domain. Following the transformation one can take advantage of the decision-directed approach in order to obtain the uncertainty of the signal,

$$
\begin{aligned}
\hat{\sigma}_{x,\text{SD}}^2(l) = &\ \beta_{\text{SD}}|\hat{X}(l-1) - \hat{A}_\mu(l-1)e^{j\hat{\alpha}_\mu(l-1)}|^2 \\
&+ (1-\beta_{\text{SD}})|Y(l) - \hat{A}_\mu(l)e^{j\hat{\alpha}_\mu(l)}|^2,
\end{aligned}
\tag{4.65}
$$

where β_{SD} is a smoothing parameter similar to (4.9). It is important to note that this estimator is phase sensitive and therefore not only taking the uncertainty in amplitude but also in spectral phase into account.

The robustness of this enhancement scheme is strongly influenced by the reliability of the estimated model parameters of speech. Therefore in McCallum and Guillemin (2013) voice activity detection is used to avoid $A_\mu \neq 0$ (implying harmonic structure of the signal model) in speech-absent analysis frames. Additional improvement can be achieved by imposing speech presence uncertainty similar to Ephraim and Malah (1984).

4.5 Experiments

By incorporating phase information into the spectral estimation of speech, the performance bounds of the conventional methods have been pushed with respect to perceived speech quality and intelligibility (see, e.g., Mowlaee and Saeidi 2013; Mowlaee and Kulmer 2015; Kulmer and Mowlaee 2015; Gerkmann and Krawczyk 2013; Gerkmann 2014a). The three experiments discussed below aim to illustrate the mechanisms that contribute to the superior performance of phase-aware solutions:

- Experiment 4.1: Proof of concept
- Experiment 4.2: Consistency
- Experiment 4.3: Sensitivity analysis.

4.5.1 Experiment 4.1: Proof of Concept

In order to illustrate how the estimators affect the processed speech signal, Figure 4.11 serves as a proof of concept. All the estimators use the exact same basic setup, meaning that the STFT setup, the *a priori* SNR estimation, and the parameters μ and β are equal for all methods (see Table 4.2). The only difference in the estimation procedures is the way they incorporate the additional phase information into the speech estimation. The phase information is provided by STFTPI Krawczyk and Gerkmann (2014). The noise-corrupted signal is obtained by mixing clean speech with white noise at a global SNR of 0 dB.

In general, the phase-aware estimators yield increased noise suppression compared to the phase-unaware baseline method in (4.23) (pictured in (c) in Figure 4.11). The rectangles highlight harmonics that are retrieved by the phase-aware estimators. The fact that the model order of the harmonic speech signal segments is not known *a priori* means that an over-harmonization takes place.

The spectral peaks resulting from the phase-unaware baseline method can be associated with musical noise, indicating that the CUP estimator (Gerkmann 2014b;

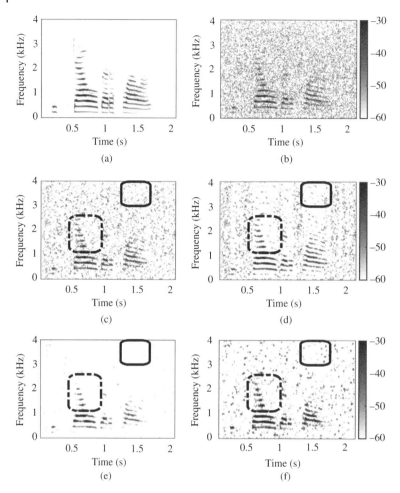

Figure 4.11 Spectrograms of (a) clean speech, (b) noise corrupted speech, (c) MMSE-STSA, (d) MMSE-STSA with given STFT phase, (e) CUP, and (f) the iterative approach.

(e) in Figure 4.11) efficiently suppresses this phenomenon. Since the complex estimator takes into account uncertainty in the phase information obtained from Krawczyk and Gerkmann (2014), the harmonic reconstruction is less aggressive than for the phase-aware MMSE-STSA estimator, fully relying on the provided phase estimate (see panel (d)). Similarly, the iterative method (panel (f)) retrieves particularly the lower frequency regions of speech very well.

4.5.2 Experiment 4.2: Consistency

In the literature it is often hypothesized that a reasonable phase estimate restores the spectro-temporal correlations in the STFT, referred to as consistency (Gerkmann *et al.* 2015). In Mowlaee and Saeidi (2013) the consistency of the enhanced STFT is used as

Table 4.2 Settings for the estimators used in the proof-of-concept experiments.

Parameter	Setting
f_s	8 kHz
μ	1
β	1
N_{DFT}	256
Frame length	32 ms
Frame shift	4 ms
β_{DD}	0.92
Noise PSD estimator	Gerkmann and Hendriks (2012)
f_0 estimator	PEFAC (Gonzalez and Brookes 2014)

Figure 4.12 Relative inconsistency for 20 randomly selected, gender balanced utterances from the TIMIT database (Garofolo *et al.* 1993) mixed at global SNRs of −5 dB, 0 dB, and 5 dB. The noise types utilized are white, babble, and factory noise. The inconsistency is normalized to (a) the outcome of the estimator in (4.33) together with STFTPI. If the amplitude of (4.33) is used together with the noisy phase for reconstruction we obtain (b). (c) is the phase-unaware baseline estimator in (4.26), and (d) is the inconsistency of the CUP estimator in (4.46).

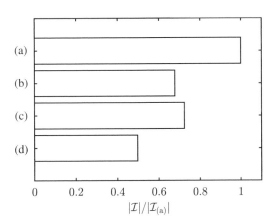

a stopping criterion, indicating whether the iterative algorithm converges or not. By evaluating the inconsistency given in (4.37) for the enhanced STFTs of each method, in Figure 4.12 we see that phase awareness leads to less inconsistent spectra than the phase-unaware solution.

4.5.3 Experiment 4.3: Sensitivity Analysis

In order to evaluate the sensitivity of the phase-aware spectral coefficient estimation solutions to the provided initial phase estimate, we are interested in the *normalized mean square error* (NMSE) of the estimated quantity. The NMSE of the amplitude estimate is given by (Ephraim and Malah 1984)

$$\text{NMSE}(A) = \frac{\mathbb{E}\left(\left(A - \hat{A}\right)^2\right)}{\sigma_x^2}. \tag{4.66}$$

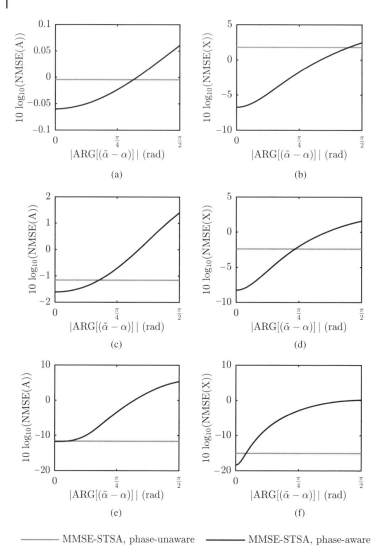

Figure 4.13 Sensitivity analysis of the phase-aware estimator in (4.34) (black curve) and its phase-unaware counterpart in (4.26) (gray curve). The left column refers to NMSE(*A*), while the right column presents NMSE(*X*). The corresponding *a priori* SNRs ξ are −15 dB in (a) and (b), 0 dB in (c) and (d), and 15 dB in (e) and (f).

If a phase estimate is used not only for amplitude estimation but also for reconstruction, the complex NMSE is of interest, which we define as

$$\text{NMSE}(X) = \frac{\mathbb{E}\left(\left(X - \hat{X}\right)(X - \hat{X})^*\right)}{\sigma_x^2}, \tag{4.67}$$

where $(\cdot)^*$ denotes the complex conjugate operator. The outcome of a Monte Carlo simulation on NMSE(*A*) and NMSE(*X*) for given *a priori* SNRs is pictured in Figure 4.13.

The simulation is based on perturbing the clean speech phase α, resulting in

$$\tilde{\alpha} = \alpha + \Delta_\alpha, \text{ with } \Delta_\alpha \in [0, \pi[. \tag{4.68}$$

Choosing the perturbation with $\Delta_\alpha \in [-\pi, 0[$ yields the same NMSEs as $\Delta_\alpha \in [0, \pi[$. The perturbed phase $\tilde{\alpha}$ is used as an estimate of the clean phase α in the two extreme-case scenarios:

1) phase-unaware amplitude estimation (4.26)
2) phase-aware amplitude estimation, assuming that the phase is given (4.34).

The realizations are sampled from distributions with known parameters μ, σ_x^2, and σ_d^2; the compression β is set to 1. From Figure 4.13, we see that an accurate initial phase estimate is indispensable in order to benefit from the additional phase information. For high SNRs the phase estimation is even disadvantageous since the required accuracy is very high. In general, the complex NMSE shows more potential than the amplitude NMSE, indicating that if an exact phase estimate is achievable it should be used for both amplitude estimation and signal reconstruction. The MATLAB® implementation for this experiment can be found in *Exp4_3.m* in the *PhaseLab Toolbox* (see the appendix for more details).

4.6   Summary

This chapter has considered the speech enhancement problem in general and in the case of available phase information. Depending on how these estimates are interpreted, different gain functions are derived. The possibilities for incorporating the additional source of information are manifold, whether for amplitude estimation and/or reconstruction. The quality of the initial phase estimate strongly influences the estimators' performance. This indicates that only reliable information should be used in this context. Under the condition that this requirement is met, the limits of phase-unaware solutions are pushed, for example in terms of musical noise suppression or reconstruction of important speech characteristics like transients and harmonic structure. Through various proof-of-concept experiments, the advantages of phase-aware processing for speech enhancement were demonstrated. As the presented phase-aware solutions all rely on the availability of an accurate estimated phase for their initialization, future studies should be directed toward finding robust phase estimators.

References

I. Andrianakis and P. R. White, Speech spectral amplitude estimators using optimally shaped gamma and chi priors, *Speech Communication*, vol. 51, no. 1, pp. 1–14, 2009.

C. Breithaupt, T. Gerkmann, and R. Martin, Cepstral smoothing of spectral filter gains for speech enhancement without musical noise, *IEEE Signal Processing Letters*, vol. 14, no. 12, pp. 1036–1039, 2007.

C. Breithaupt, T. Gerkmann, and R. Martin, *A Novel A Priori SNR Estimation Approach Based on Selective Cepstro-Temporal Smoothing*, Proceedings of the IEEE International

Conference on Acoustics, Speech and Signal Processing (ICASSP), pp. 4897–4900, 2008a.

C. Breithaupt, M. Krawczyk, and R. Martin, *Parameterized MMSE Spectral Magnitude Estimation for the Enhancement of Noisy Speech*, Proceedings of the IEEE International Conference on Acoustics, Speech and Signal Processing (ICASSP), pp. 4037–4040, 2008b.

O. Cappé, Elimination of the musical noise phenomenon with the Ephraim and Malah noise suppressor, *IEEE Transactions on Speech and Audio Processing*, vol. 2, no. 2, pp. 345–349, 1994.

Y. Ephraim and D. Malah, *Speech Enhancement Using Vector Spectral Subtraction Amplitude Estimation*, Proc. 13th IEEE Convention of Electrical & Electronics Engineers, 1983.

Y. Ephraim and D. Malah, Speech enhancement using a minimum mean-square error short-time spectral amplitude estimator, *IEEE Transactions on Acoustics, Speech and Signal Processing*, vol. 32, no. 6, pp. 1109–1121, 1984.

Y. Ephraim and D. Malah, Speech enhancement using a minimum mean-square error log-spectral amplitude estimator, *IEEE Transactions on Acoustics, Speech and Signal Processing*, vol. 33, no. 2, pp. 443–445, 1985.

J. S. Erkelens, R. C. Hendriks, R. Heusdens, and J. Jensen, Minimum mean-square error estimation of discrete Fourier coefficients with generalized gamma priors, *IEEE Transactions on Audio, Speech, and Language Processing*, vol. 15, no. 6, pp. 1741–1752, 2007.

J. S. Garofolo, L. F. Lamel, W. M. Fisher, J. G. Fiscus, D. S. Pallett, and N. L. Dahlgren, *DARPA TIMIT Acoustic Phonetic Continuous Speech Corpus CDROM*, NIST, 1993.

T. Gerkmann and R. Hendriks, Unbiased MMSE-based noise power estimation with low complexity and low tracking delay, *IEEE Transactions on Audio, Speech, and Language Processing*, vol. 20, no. 4, pp. 1383–1393, 2012.

T. Gerkmann and M. Krawczyk, MMSE-optimal spectral amplitude estimation given the STFT-phase, *IEEE Signal Processing Letters*, vol. 20, no. 2, pp. 129–132, 2013.

T. Gerkmann, *MMSE-Optimal Enhancement of Complex Speech Coefficients with Uncertain Prior Knowledge of the Clean Speech Phase*, Proceedings of the IEEE International Conference on Acoustics, Speech and Signal Processing (ICASSP), pp. 4478–4482, 2014a.

T. Gerkmann, Bayesian estimation of clean speech spectral coefficients given *a priori* knowledge of the phase, *IEEE Transactions on Signal Processing*, vol. 62, no. 16, pp. 4199–4208, 2014b.

T. Gerkmann, M. Krawczyk-Becker, and J. Le Roux, Phase processing for single-channel speech enhancement: History and recent advances, *IEEE Signal Processing Magazine*, vol. 32, no. 2, pp. 55–66, 2015.

S. Gonzalez and M. Brookes, PEFAC: A pitch estimation algorithm robust to high levels of noise, *IEEE/ACM Transactions on Audio, Speech, and Language Processing*, vol. 22, no. 2, pp. 518–530, 2014.

I. S. Gradshteyn and I. M. Ryzhik, *Table of Integrals, Series, and Products*, 7th edn. Elsevier/Academic Press, 2007.

D. Griffin and J. Lim, Signal estimation from modified short-time Fourier transform, *IEEE Transactions on Acoustics, Speech and Signal Processing*, vol. 32, no. 2, pp. 236–243, 1984.

R. Hendriks, R. Heusdens, and J. Jensen, An MMSE estimator for speech enhancement under a combined stochastic–deterministic speech model, *Audio, Speech, and Language Processing, IEEE Transactions on*, vol. 15, no. 2, pp. 406–415, 2007.

R. Hendriks, R. Heusdens, and J. Jensen, *MMSE-Based Noise PSD Tracking with Low Complexity*, Proceedings of the IEEE International Conference on Acoustics Speech and Signal Processing (ICASSP), pp. 4266–4269, 2010.

S. M. Kay, *Fundamentals of Statistical Signal Processing, Volume I: Estimation Theory*, Prentice Hall, 1993.

M. Krawczyk and T. Gerkmann, *STFT Phase Improvement for Single Channel Speech Enhancement*, Proceedings of the International Workshop on Acoustic Signal Enhancement (IWAENC), pp. 1–4, 2012.

M. Krawczyk, R. Rehr, and T. Gerkmann, *Phase-Sensitive Real-Time Capable Speech Enhancement under Voiced-Unvoiced Uncertainty*, Proceedings of the European Signal Processing Conference (EUSIPCO), pp. 1–5, 2013.

M. Krawczyk and T. Gerkmann, STFT phase reconstruction in voiced speech for an improved single-channel speech enhancement, *IEEE/ACM Transactions on Audio, Speech, and Language Processing*, vol. 22, no. 12, pp. 1931–1940, 2014.

J. Kulmer and P. Mowlaee, Phase estimation in single channel speech enhancement using phase decomposition, *IEEE Signal Processing Letters*, vol. 22, no. 5, pp. 598–602, 2015.

J. Lim and A. Oppenheim, Enhancement and bandwidth compression of noisy speech, *Proceedings of the the IEEE*, vol. 67, no. 12, pp. 1586–1604, 1979.

P. C. Loizou and G. Kim, Reasons why current speech-enhancement algorithms do not improve speech intelligibility and suggested solutions, *IEEE Transactions on Audio, Speech, and Language Processing*, vol. 19, no. 1, pp. 47–56, 2011.

T. Lotter and P. Vary, *Noise Reduction by Joint Maximum A Posteriori Spectral Amplitude and Phase Estimation with Super-Gaussian Speech Modelling*, Proceedings of the European Signal Processing Conference (EUSIPCO), pp. 1457–1460, 2004.

R. McAulay and M. Malpass, Speech enhancement using a soft-decision noise suppression filter, *IEEE Transactions on Acoustics, Speech and Signal Processing*, vol. 28, no. 2, pp. 137–145, 1980.

M. McCallum and B. Guillemin, Stochastic–deterministic MMSE STFT speech enhancement with general a priori information, *IEEE Transactions on Audio, Speech, and Language Processing*, vol. 21, no. 7, pp. 1445–1457, 2013.

D. Malah, R. Cox, and A. Accardi, *Tracking Speech-Presence Uncertainty to Improve Speech Enhancement in Non-Stationary Noise Environments*, Proceedings of the IEEE International Conference on Acoustics, Speech, and Signal Processing (ICASSP), pp. 789–792, vol. 2, 1999.

R. Martin, *Spectral Subtraction Based on Minimum Statistics*, Proceedings of the European Signal Processing Conference (EUSIPCO), pp. 1182–1185, 1994.

R. Martin, Noise power spectral density estimation based on optimal smoothing and minimum statistics, *IEEE Transactions on Speech and Audio Processing*, vol. 9, no. 5, pp. 504–512, 2001.

R. Martin, Speech enhancement based on minimum mean-square error estimation and supergaussian priors, *IEEE Transactions on Speech and Audio Processing*, vol. 13, no. 5, pp. 845–856, 2005.

R. Martin, Bias compensation methods for minimum statistics noise power spectral density estimation, *Signal Processing*, vol. 86, no. 6, pp. 1215–1229, 2006.

P. Mowlaee, R. Saeidi, and R. Martin, *Phase Estimation for Signal Reconstruction in Single-Channel Source Separation*, Proceedings of the International Conference on Spoken Language Processing (INTERSPEECH), pp. 1548–1551, 2012.

P. Mowlaee and R. Saeidi, Iterative closed-loop phase-aware single-channel speech enhancement, *IEEE Signal Processing Letters*, vol. 20, no. 12, pp. 1235–1239, 2013.

P. Mowlaee and J. Kulmer, Phase estimation in single-channel speech enhancement: Limits-potential, *IEEE/ACM Transactions on Audio, Speech, and Language Processing*, vol. 23, no. 8, pp. 1283–1294, 2015.

K. Paliwal, K. Wójcicki, and B. J. Shannon, The importance of phase in speech enhancement, *Speech Communication*, vol. 53, no. 4, pp. 465–494, 2011.

D. Thomson, Spectrum estimation and harmonic analysis, *Proceedings of the IEEE*, vol. 70, no. 9, pp. 1055–1096, 1982.

S. Vanambathina and T. K. Kumar, Speech enhancement by Bayesian estimation of clean speech modeled as super Gaussian given a priori knowledge of phase, *Speech Communication*, vol. 77, pp. 8–27, 2016.

P. Vary, Noise suppression by spectral magnitude estimation mechanism and theoretical limits, *Signal Processing*, vol. 8, no. 4, pp. 387–400, 1985.

D. Wang and J. Lim, The unimportance of phase in speech enhancement, *IEEE Transactions on Acoustics, Speech and Signal Processing*, vol. 30, no. 4, pp. 679–681, 1982.

C. H. You, S. N. Koh, and S. Rahardja, β-order MMSE spectral amplitude estimation for speech enhancement, *IEEE Transactions on Speech and Audio Processing*, vol. 13, no. 4, pp. 475–486, 2005.

5

Phase Processing for Single-Channel Source Separation

Pejman Mowlaee and Florian Mayer

Graz University of Technology, Graz, Austria

5.1 Chapter Organization

In this chapter, we provide another practical application to demonstrate the importance of phase processing in speech communication. In contrast to Chapter 4, which is focused on speech enhancement, this chapter will examine the role of phase in *single-channel source separation* (SCSS). After a review of conventional phase-unaware SCSS methods, we proceed with several examples where phase processing has been incorporated for SCSS. We highlight the contribution of phase processing in two steps: spectral modification and signal reconstruction. The recent advances with regard to incorporating phase in SCSS for the signal interaction function, phase estimation for signal reconstruction, time–frequency masking, and matrix factorization for spectral modification will be thoroughly demonstrated. The chapter supplies various examples to visualize and quantify how phase processing pushes the achievable performance of the state-of-the-art phase-unaware SCSS methods.

5.2 Why Single-Channel Source Separation?

5.2.1 Background

We often encounter situations where simultaneous acoustic events occur, among which we might be interested only in one particular signal. For instance, imagine a busy intersection with car noise as well as an acoustic alarm warning or, as an alternative scenario, a room crowded with people. Cherry termed such an adverse noise scenario as the *cocktail party problem*: "How do we recognize what one person is saying when others are speaking at the same time (the 'cocktail party problem')? On what logical basis could one design a machine (filter) for carrying out such an operation?" (Cherry 1953). In all such scenarios, too many acoustic events could overload our perception and distract our focus on the desired acoustic event. People with normal hearing are able to concentrate on one acoustic event at a time by neglecting the interfering ones. This is automatically

Single Channel Phase-Aware Signal Processing in Speech Communication: Theory and Practice, First Edition.
Pejman Mowlaee, Josef Kulmer, Johannes Stahl, and Florian Mayer.
© 2017 John Wiley & Sons, Ltd. Published 2017 by John Wiley & Sons, Ltd.

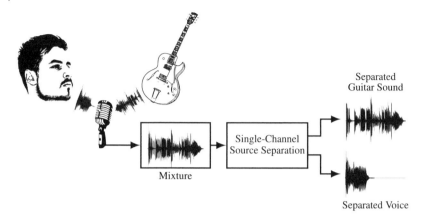

Figure 5.1 A general scenario with mixed sources: A voice is masked by a guitar sound in the background, both recorded with one microphone. SCSS is capable of separating both underlying sources from their mixture.

performed by the human auditory system. In contrast, hearing-impaired people not only suffer from loss of intelligibility, but also forfeit their ability to separate acoustic events in adverse noise conditions. All these examples emphasize the importance of a *single-channel source separation* (SCSS) processor for improved machine listening or *automatic speech recognition* (ASR) performance.

   While the human auditory system performs a decent job in identifying and separating the individual sources, machines fall short in their achievable separation performance. As a consequence, the resulting automatic speech recognition performance decreases. Figure 5.1 illustrates a typical SCSS scenario, where the mixture of two sound sources is recorded by only one microphone. Several source separation and recognition challenges, for example the *signal separation evaluation campaign* (SiSEC) (Araki *et al.* 2012) or CHiME (Barker *et al.* 2013; Vincent *et al.* 2013; Barker *et al.* 2015), have been regularly organized as an attempt to test the performance of SCSS methods in a realistic scenario by their resulting ASR accuracy. Realistic scenarios, such as reverberant and multi-source mixtures, were taken into account in order to evaluate the source separation performance.

5.2.2 Problem Formulation

Let $y(n) = \sum_{q=1}^{Q} x_q(n)$ be an audio mixture composed of Q audio sources where $x_q(n)$ represents the qth underlying source with $q \in [1, Q]$. Let $\hat{x}_q(n)$ denote the estimated separated signals from the mixture. Taking the Fourier transform of overlapping signal segments, we obtain $X_q(k, l)$ and $Y(k, l)$ as the qth source and the mixture STFT, and $\phi_q(k, l)$ and $\phi_y(k, l)$ as the qth source phase and mixture phase spectra, respectively. The mixture spectrum $Y(k, l)$, at each frequency bin k and frame l, is the result of the vector addition of the underlying sources. Figure 5.2 shows the vector diagram for the SCSS problem. Applying the Fourier transformation, we obtain

$$|Y(k, l)|e^{j\phi_y(k,l)} = \sum_{q=1}^{Q} |X_q(k, l)|e^{j\phi_q(k,l)}. \tag{5.1}$$

Figure 5.2 Geometry of the SCSS problem.

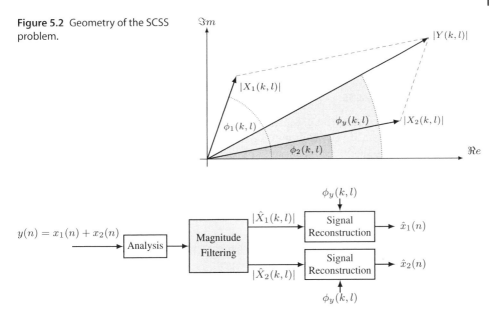

Figure 5.3 Conventional SCSS principle using mixture phase at signal reconstruction stage.

In the course of this chapter, we define $|\hat{X}_q(k,l)|$ and $\hat{\phi}_q(k,l)$ as the estimated magnitude and phase spectra of the qth source. In order to recover both sources entirely, the magnitude and phase information have to be estimated from the single-channel recorded mixture. Therefore, the problem of separating Q sources from one observation given in (5.1) is an ill-conditioned one, since there is only one equation for $Q > 1$ unknowns.

Figure 5.3 shows the block diagram of a conventional SCSS method consisting of two steps: (i) magnitude filtering, and (ii) signal reconstruction. With regards to phase processing, in this chapter we will address two aspects:

- How does the knowledge of phase information contribute to improve the signal reconstruction of the separated signals?
- How does phase information contribute to any further improvement of the separation accuracy of the source magnitude spectra?

5.3 Conventional Single-Channel Source Separation

Conventional phase-unaware SCSS algorithms aim at estimating the source magnitude spectra. The estimation outcome for the separated signal $\hat{x}_q(n)$ is obtained by combining the mixture phase with the qth magnitude spectrum. In particular, a filter is designed and applied to the magnitude spectrum of the mixed signal, while the spectral phase of the mixed signal is directly used for signal reconstruction. Commonly used SCSS algorithms follow two global approaches:

- source-driven
- model-driven.

Both approaches employ the spectral amplitude information only, with no implicit usage of the phase information, either at the amplitude estimation stage or at the signal reconstruction stage. In the following, we present a review of each group and explain how phase information is neglected in these methods.

5.3.1 Source-Driven SCSS

Bregman (1990) described the process by which humans separate each desired acoustic event out of a real-life environment, called *auditory scene analysis* (ASA). This is a challenging task, as the human ears only have access to the mixture of all underlying acoustic events, though our brain prioritizes or neglects certain sound events. *Computational ASA (CASA)* can be considered as a computerized listener trying to imitate this human behavior. To allocate the underlying acoustic events, CASA takes no prior knowledge into account. However, the separated signals rely on several signal attributes including onsets, offsets, and pitch trajectories. The scene analysis is performed in two stages (see Figure 5.4):

- **Segmentation:** At this stage, the mixture signal is decomposed into individual segments by analyzing several attributes like pitch trajectories, common onset/offset, harmonicity, and periodicity. To segment individual sources, the mixture signal is transformed into time–frequency components, where adjacent time–frequency fractions are subsequently formed into a segment. The goal is to collect the time–frequency information that can be linked to the same source.
- **Grouping:** Identified segments that are likely to originate from the same source are grouped and distributed to an auditory stream. Grouping consists of two stages: simultaneous and sequential organization. The simultaneous organization groups segments across frequency, mostly relying on the pitch information commonly provided by a multi-pitch estimator (Hu and Wang 2004; Wohlmayr and Pernkopf 2013). Therefore, voiced segments are grouped into simultaneous streams. Sequential grouping of the ascertained simultaneous streams and unvoiced parts is performed to create streams linked to the individual speakers. Hu and Wang (2004) proposed segmentation in combination with sequential grouping, which led to a robust grouping performance and improved SNR of segregated speech.

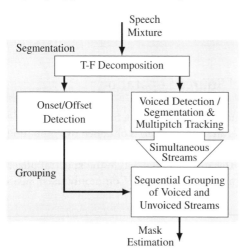

Figure 5.4 Block diagram of a CASA system inspired by Wang (2005), comprised of segmentation and grouping stages.

After collecting information about an auditory scene, CASA utilizes this knowledge to estimate the *ideal binary mask* (IBM) and *ideal ratio mask* (IRM). Each estimated IBM and IRM is eventually applied to the mixture to obtain an estimate of the underlying sources.

5.3.1.1 Ideal Binary Mask

The ideal binary mask is considered as the ultimate goal of CASA (Wang 2005). Time–frequency regions of the target signal stronger than the interference are preserved, while the regions weaker than the interfering signal are discarded. Taking the Fourier transform, the ideal binary mask is defined as:

$$\text{IBM}(k,l) = \begin{cases} 1 & \text{SSR}(k,l) > \text{LC}, \\ 0 & \text{otherwise.} \end{cases} \tag{5.2}$$

where $\text{SSR}(k,l) = \frac{|X_1(k,l)|^2}{|X_2(k,l)|^2}$ is called the *signal-to-signal ratio* (SSR), defined as the power ratio of target to masker. It is important to note that in source separation literature, the SSR term is often used rather than the SNR used in speech enhancement literature. The *local criterion* (LC) is the threshold that determines the boundary at which the binary mask is either 1 or 0. Typically, an SSR of 0 dB is used as a masking criterion. The ideal binary mask $\text{IBM}(k,l)$ is applied to the mixture spectrum to obtain the target signal. Employing an IBM on a mixture signal improves the intelligibility of the desired signal (Wang 2005; Li and Wang 2008) as well as the ASR performance (Srinivasan *et al.* 2006), although IBM is reported to have poor perceived quality. Williamson *et al.* (2014) presented an overview of different techniques to improve the perceived quality of IBM for SCSS.

5.3.1.2 Ideal Ratio Mask

Wang *et al.* (2014) showed that, compared to IBM, IRM achieves better separation results. It is defined as

$$\text{IRM}(k,l) = \left(\frac{|X_1(k,l)|^2}{|X_1(k,l)|^2 + |X_2(k,l)|^2} \right)^o, \tag{5.3}$$

where $|X_1(k,l)|^2$ and $|X_2(k,l)|^2$ denote the power of the underlying target and masking speakers, respectively, and o is a tuning factor used to compress the mask. The authors determined o = 0.5 as the optimal choice, which leads (5.3) to be equivalent to a square-root Wiener filter. Similar to IBM, the estimated IRM is applied to the mixture signal to obtain the desired target or masking signal.

Figure 5.5 illustrates the outcomes of IBM and IRM according to the local SSR $\left(\text{SSR}_{\text{local}} = \frac{|X_1(k,l)|^2}{|X_2(k,l)|^2} \right)$ for several numbers of frames at frequency bin $k = 6$. The mixture is composed of one male and one female utterance mixed at 0 dB global SSR. The time frequency representations achieved for both IBM and IRM masks reveal that IRM preserves the energy level at time–frequency cells for the desired speaker.

5.3.2 Model-Based SCSS

Model-driven methods train dictionaries of individual source types in order to perform a dictionary-based separation. In contrast to their source-driven counterparts,

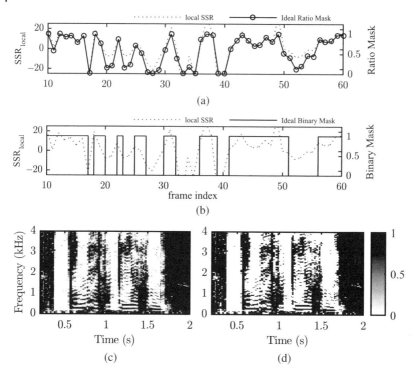

Figure 5.5 Computation of the local SSR for the target source for (a) ideal ratio mask ($\beta = 1$) and (b) ideal binary mask, at frequency bin $k = 6$. Below, the time frequency representation of the IRM ($\beta = 1$) (c) and IBM (d) are shown, respectively.

model-based methods show a more accurate source separation performance, albeit they are computationally more expensive than CASA methods. In the following, we present a brief description of model-based methods and identify two important groups: deep learning and *non-negative matrix factorization* (NMF).

Pre-trained dictionaries for the underlying sources in the mixture are applied to separate them from the single-channel mixture. Compared to the source-driven group, these methods provide a more accurate source separation performance, although at the cost of greater computational complexity. For example, Hershey *et al.* (2010) demonstrated that for a small vocabulary size corpus (like GRID; Cooke *et al.* 2006), it is possible to even outperform the human listener word recognition accuracy; hence, their system was called *super-human multi-talker speech recognition*. The results were, however, restricted to the limited grammar used in the transcription of the database. For a general review of model-based SCSS methods see Cooke *et al.* (2010), Mowlaee (2010), Hershey *et al.* (2013), and Virtanen *et al.* (2015).

In the following, we briefly review two important members of the model-based group for SCSS: deep learning and NMF. In the course of this chapter, after a brief introduction on recent advances made in deep learning methods for SCSS, we will proceed with a deeper focus on NMF and its phase-aware extensions, as discussed in Section 5.4.1.

5.3.2.1 Deep Learning

Recently, *deep neural networks* (DNN) have been employed for supervised SCSS. The bottom line is that the DNN learns the mapping from the noisy signal to the corresponding clean version using a mapping function as an estimate for the time–frequency mask function (Jin and Wang 2009; Wang and Wang 2013). The training target used for DNNs is set to an ideal time–frequency mask, e.g., IBM or IRM. For the IBM scenario, the SCSS degenerates to a binary classification problem, which is a known machine learning task. While the IBM improves the intelligibility, the use of IRM was shown to improve both speech intelligibility and perceived quality of the separated output signals (Wang *et al.* 2014).

In general, two groups of methods are distinguished, as shown in Figure 5.6:

- direct learning
- indirect learning.

Both methods aim to find an ideal time–frequency mask $G(k, l)$ given a mixed signal $Y(k, l)$. For *direct learning*, the deep model is used either as a supervised learning algorithm (Wang and Wang 2013) or as a structured prediction (Wang and Wang 2012) to estimate the IRM/IBM. Several neural network architectures, including multi-layer perceptron, deep belief network, and general stochastic network (Zöhrer *et al.* 2015), along with different time–frequency masks (Wang and Wang 2013) have been studied. For a full review of supervised speech separation methods using DNN, see Wang (2015). Studies showed that the cosine of the phase deviation ($\cos(\phi_{\text{dev}}(k, l))$, as defined in Chapter 1) applied to IRM is advantageous (Erdogan *et al.* 2015; Weninger *et al.* 2015). Also, the application of two neural nets, one for the time–frequency mask and another for the inverse Fourier transform (Wang and Wang 2015) has been reported successful for estimating time domain signal. Theses studies encourage the incorporation of an enhanced phase spectrum in the framework of deep learning for SCSS (Wang 2015). We will provide an overview of time–frequency masks in Section 5.4.3.

For *indirect learning*, the ideal time–frequency mask estimation relies on employing two separate models that are both trained on source-specific data. At the separation stage, a combination of the two models is used to estimate the desired mask required to retrieve the underlying speakers. Examples are SCSS methods based on

Figure 5.6 Different approaches in deep learning inspired by Zöhrer *et al.* (2015) using (a) one model M_1 directly, (b) indirect learning of the ideal time frequency mask using two models M_1 and M_2. The models learn the time frequency mask $G(k, l)$ used to separate sources from mixture $Y(k, l)$.

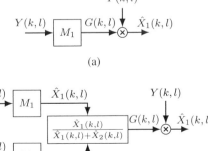

vector quantization, Gaussian mixture models, graphical models, and hidden Markov models (Roweis 2003; Mowlaee *et al.* 2012b; Hershey *et al.* 2010).

Providing a detailed review of deep models for SCSS is not within the scope of this book. In the rest of this chapter, therefore, we focus only on the SCSS methods that incorporate phase-aware signal processing for improved signal reconstruction or improved estimation of the source magnitude spectra.

5.3.2.2 Non-Negative Matrix Factorization

The so-called NMF has become increasingly popular for solving the SCSS problem. The expression *non-negative* refers to the magnitude spectrum of a signal, which is always positive.

Let $V_q(m, l)$ be the activation function of source q for atom m and frame l with corresponding activation vector $\mathbf{v}_{m,q} \in \mathbb{R}_{\geq 0}^{L \times 1}$ as $\mathbf{v}_{m,q} = [V_q(m, 1), \cdots, V_q(m, L)]^T$ so that $\mathbf{V}_q = [\mathbf{v}_{1,q} \cdots \mathbf{v}_{M,q}]^T$ be the so-called *activation matrix* of dimension $M \times L$, M being the total number of atoms and L the total number of frames. The corresponding matrix $\mathbf{B}_q = [\mathbf{b}_{1,q} \cdots \mathbf{b}_{M,q}]$ is called the *basis matrix* of dimension $K \times M$ with K the number of filters in the filterbank (the number of DFT bins of the STFT), whose columns are basis vectors of the mth atom and qth source denoted by $\mathbf{b}_{m,q} = [B_q(1, m), \cdots, B_q(K, m)]^T$. The composition of $|\mathbf{X}_q|$ in terms of the basis and activations is given by

$$|\mathbf{X}_q| \approx \mathbf{B}_q \mathbf{V}_q. \tag{5.4}$$

The NMF approach aims to approximately factorize a given $K \times L$ non-negative data matrix $|\mathbf{X}_q|$ into a $K \times M$ non-negative basis matrix $\mathbf{B}_q$, and $M \times L$ non-negative activation matrix $\mathbf{V}_q$. NMF is then expressed as follows:

$$
\begin{aligned}
\text{Given:} &\quad |\mathbf{X}_q| \in \mathbb{R}_{\geq 0}^{K \times L} \quad \text{and} \quad K \in \mathbb{N}_{>0}, \\
\text{factorize:} &\quad |\mathbf{X}_q| \approx |\hat{\mathbf{X}}_q| = \mathbf{B}_q \mathbf{V}_q, \\
\text{subject to:} &\quad \mathbf{V}_q \in \mathbb{R}_{\geq 0}^{M \times L} \quad \text{and} \quad \mathbf{B}_q \in \mathbb{R}_{\geq 0}^{K \times M}.
\end{aligned}
\tag{5.5}
$$

Figure 5.7 illustrates how NMF approximates a speech signal composed of M atoms.

To accomplish the signal decomposition targeted in NMF, we need to find the $\mathbf{B}_q$ and $\mathbf{V}_q$ which most closely jointly satisfy (5.4). To this end, we stack $Y(k, l)$ for all k and l to the matrix $\mathbf{Y}$ and the divergence between $|\mathbf{Y}|$ and its approximated version using the signal decompositions $|\hat{\mathbf{Y}}| = \sum_{q=1}^{Q} \mathbf{B}_q \mathbf{V}_q$ is considered:

$$\text{Div}(|\mathbf{Y}| \| |\hat{\mathbf{Y}}|) = \sum_{k,l} d(y_{k,l,q}, \hat{y}_{k,l,q}), \tag{5.6}$$

where $y_{k,l,q}$ and $\hat{y}_{k,l,q}$ are the (k, l)th entries of $|\mathbf{Y}_q|$ and $|\hat{\mathbf{Y}}_q|$, and $d(\cdot)$ is the divergence between two scalar inputs. The optimal values $\mathbf{B}_q^*$ and $\mathbf{V}_q^*$ of $\mathbf{B}_q$ and $\mathbf{V}_q$ are obtained by minimizing the divergence in (5.6), and we have:

$$\mathbf{B}_q^*, \mathbf{V}_q^* = \arg\min_{\mathbf{B}_q, \mathbf{V}_q} \quad \text{Div}(|\mathbf{Y}| \| |\hat{\mathbf{Y}}|) \quad \mathbf{B}_q \geq 0, \mathbf{V}_q \geq 0. \tag{5.7}$$

The optimization defined in (5.7) emphasizes that both, the optimal atoms $\mathbf{V}_q^*$ and optimal basis $\mathbf{B}_q^*$ are simultaneously required for decomposition.

Several variants of NMF, taking sparsity and temporal dynamicity constraints into account, have been successfully applied for SCSS, for example Virtanen (2007),

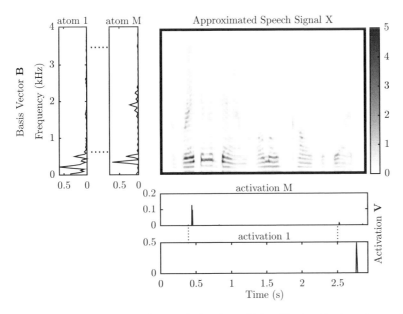

Figure 5.7 Decomposition of a speech signal, combining *M* trained basis vectors and estimated activations.

Smaragdis (2007), Raj *et al.* (2011), Mohammadiha *et al.* (2013), and Virtanen *et al.* (2015). The three most common measures in (5.6) are the Frobenius norm, the Kullback–Leibler divergence (KL), and the Itakura–Saito distance. For an overview, see Fevotte and Idier (2011) and Virtanen *et al.* (2015).

The optimization problem defined in (5.7) is not jointly convex, but biconvex in $\mathbf{V}_q$ and $\mathbf{B}_q$, which means if either one of the two variables is known it is convex in the remaining variable. If we assume that, for example, the atoms $\mathbf{B}_q$ are known, then the decomposition degenerates to only estimating the activations $\mathbf{V}_q$ and we have:

$$\mathbf{V}_q^* = \arg\min_{\mathbf{V}_q} \quad \mathrm{Div}(|\mathbf{Y}| \,||\, |\hat{\mathbf{Y}}|) \quad \mathbf{V}_q \geq 0. \tag{5.8}$$

In a similar way, $\mathbf{B}_q^*$ could be found when $\mathbf{V}_q$ is pre-specified. Using multiplicative updates to decrease the KL-divergence, the activations and basis estimates are iteratively obtained as

$$\mathbf{B}_q \leftarrow \mathbf{B}_q \otimes \frac{\frac{|\mathbf{X}_q|}{\mathbf{B}_q \mathbf{V}_q} \mathbf{V}_q^T}{\mathbf{1}\mathbf{V}_q^T},$$

$$\mathbf{V}_q \leftarrow \mathbf{V}_q \otimes \frac{\mathbf{B}_q^T \frac{|\mathbf{X}_q|}{\mathbf{B}_q \mathbf{V}_q}}{\mathbf{B}_q^T \mathbf{1}}, \tag{5.9}$$

where all divisions are element-wise, $\otimes$ is an element-wise matrix product, and $\mathbf{1}$ is the all-ones matrix of the same dimension as $|\mathbf{X}_q|$. The procedure of re-estimation of bases and weights is illustrated in Figure 5.8.

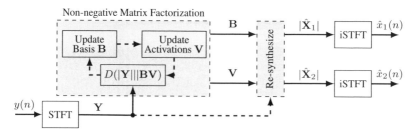

Figure 5.8 Multiplicative update for the basis matrix **B** and activation matrix **V** in NMF to approximate the underlying source magnitude spectrum **Y**.

It is important to note that for signal reconstruction in NMF, the estimated compositional parts of the sources are used to calculate a soft mask to be applied on the mixture

$$|\hat{\mathbf{X}}_q| = |\mathbf{Y}| \otimes \frac{\hat{\mathbf{B}}_q \hat{\mathbf{V}}_q}{\sum_{q=1}^{Q} \hat{\mathbf{B}}_q \hat{\mathbf{V}}_q}. \tag{5.10}$$

The filter, similar to the Wiener filter, ensures that the summation of the separated sources is equal to the mixture signal (Virtanen *et al.* 2015). Given the complex spectral estimate composed of mixture phase and estimated magnitude spectrum, an inverse Fourier transform and overlap-and-add routine is applied to reconstruct the separated time domain signal $\hat{x}_q(n)$.

5.4 Phase Processing for Single-Channel Source Separation

From the general AMS framework applied to the SCSS task, there are three ways to incorporate the estimated information about the clean spectral phase into the solution of the SCSS problem:

- Replacing the mixture phase with a clean phase estimate for signal reconstruction;
- Using the spectral phase to achieve a phase-aware spectral magnitude estimate;
- Jointly estimating the spectral magnitude and phase.

Within the rest of this chapter, we will detail the following four distinguished phase-aware processing approaches found in the SCSS literature:

- *complex matrix factorization* (CMF)
- phase estimation for signal reconstruction
- phase-aware time–frequency masks
- phase-sensitive signal interaction model.

5.4.1 Complex Matrix Factorization Methods

5.4.1.1 Complex Matrix Factorization

The method of *non-negative matrix factorization* (NMF) presented in Section 5.3.2.2 separates the source magnitude spectra only while the mixture phase is used directly for signal reconstruction. Kameoka *et al.* (2009) extended the idea of NMF to estimate the sources in the complex domain, termed *complex matrix factorization* (CMF):

$$\hat{X}_q(k, l) = \sum_{m=1}^{M} B_q(k, m) V_q(m, l) e^{j\phi_m(k,l)}, \tag{5.11}$$

where the factorization now incorporates the missing phase information $e^{j\phi_m(k,l)}$ at each time–frequency point for each atom m. Let $\Theta = \{\mathbf{B}, \mathbf{V}, \mathbf{\Phi}\}$, where $\mathbf{B}$ and $\mathbf{V}$ are similar to NMF while $\mathbf{\Phi}$ is a three-dimensional matrix of size $K \times L \times M$. The optimal values of Θ are given by solving the following optimization problem:

$$\text{Given:} \quad \mathbf{X}_q \in \mathbb{C}^{K \times L} \quad \text{and} \quad K \in \mathbb{R}_{>0},$$

$$\text{factorize:} \quad \mathbf{X}_q \approx \hat{\mathbf{X}}_q = \sum_{m=1}^{M} B_q(k, m) V_q(m, l) e^{j\phi_m(k,l)},$$

$$\text{subject to:} \quad \sum_{k} B(k, m) = 1 \quad (\forall m = 1, \dots, M),$$

$$\mathbf{V}_q \in \mathbb{R}_{\geq 0}^{M \times L}, \mathbf{B}_q \in \mathbb{R}_{\geq 0}^{K \times M}, \mathbf{\Phi}_q \in \mathbb{R}^{K \times L \times M}. \tag{5.12}$$

It is important to note that the method in Kameoka *et al.* (2009) was originally called *complex non-negative matrix factorization* since the inclusion of the phase term prevents it from being expressed as a true factorization while both bases and weights still contain non-negative elements. However, in this chapter, due to the similarity of this method and NMF, and to disambiguate between the complex and non-negative methods, we will refer to this algorithm as *complex matrix factorization*.

The CMF algorithm in Kameoka *et al.* (2009) follows these steps:

1) The basis matrix $\mathbf{B}_q$ and activations $\mathbf{V}_q$ are initialized with values close to 1, while the phase is initialized with the phase of the mixture.
2) Compute

$$\beta_q = \frac{\mathbf{B}_q \mathbf{V}_q}{\sum_{q=1}^{Q} \mathbf{B}_q \mathbf{V}_q}, \tag{5.13}$$

where division is element-wise.
3) Compute

$$\overline{X}_q(k, l) = \sum_{m=1}^{M} B_q(k, m) V_q(m, l) e^{j\phi_m(k,l)} + \beta_q(k, l)(X(k, l) - \hat{X}(k, l)). \tag{5.14}$$

4) Update $\mathbf{V}_q$:

$$\overline{\mathbf{V}}_q = \mathbf{V}_q. \tag{5.15}$$

5) Compute

$$\mathbf{\Phi}_q = \angle \frac{\mathbf{X}_q}{|\mathbf{X}_q|}, \tag{5.16}$$

the basis matrix

$$B_q(k, m) = \frac{\sum_{l=1}^{L} \frac{V_q(m,l)}{\beta_q(k,l)} \text{Re} \left(\overline{X}_q(k, l) e^{j\phi_m(k,l)}\right)}{\sum_{l=1}^{L} \frac{V_q^2(m,l)}{\beta_q(k,l)}}, \tag{5.17}$$

and the activation matrix

$$V_q(m, l) = \frac{\sum_{k=1}^{K} \frac{B_q(k,m)}{\beta_q(k,l)} \text{Re} \left(\overline{X}_q(k, l) e^{j\phi_m(k,l)} \right)}{\sum_{k=1}^{K} \frac{B_q^2(k,m)}{\beta_q(k,l)} + \lambda \rho_0 |\overline{V}_q(m, l)|^{\rho_0 - 2}}, \tag{5.18}$$

with λ used to promote a sparsity constraint applied on the activations and ρ_0 selecting ρ_0-sparsity, with $0 < \rho_0 < 2$.

The iterations continue until the change in the error from one iteration to the next converges. The separated sources are then resynthesized by combining the estimated basis, activations, and phase information obtained by (5.11). Compared to NMF, CMF (Kameoka *et al.* 2009) estimates the phase for each basis matrix to fit the sum of the estimated sources to the mixture. The CMF algorithm has a very high dimensionality for signal phase values, which has a negative effect on the computational load of the algorithm due to over-parametrization.

5.4.1.2 Complex Matrix Factorization with Intra-Source Additivity

Although CMF (Kameoka *et al.* 2009) outperforms NMF in terms of source separation measures, it also introduces over-parametrization. Therefore, King (2012) proposed the *complex matrix factorization with intra-source additivity* (CMFWISA), where the signal model was changed to

$$X_q(k, l) = \sum_{m=1}^{M} B_q(k, m) V_q(m, l) e^{j\phi_q(k,l)}, \tag{5.19}$$

where the individual phase value of each source q denoted by $\phi_q(k, l)$ is taken into account. Using multiplicative updates, estimates of $\mathbf{B}_q$ and $\mathbf{V}_q$ are used to reconstruct the estimated separated outputs. King (2012) studied two different possibilities for signal reconstruction, constrained on the *signal-to-interference ratio* (SIR) or on the *signal-to-artifact ratio* (SAR):

- *maximizing SIR synthesis* (SIRMAX), where the resynthesized source provides more suppression of interfering signals,

$$\hat{X}_q^{\text{SIRMAX}}(k, l) = \sum_{m=1}^{M} B_q(k, m) V_q(m, l) e^{j\phi_q(k,l)}. \tag{5.20}$$

- *maximizing SAR synthesis* (SARMAX), where the artifacts are minimized at the expense of lower suppression of the masker signal:

$$\hat{X}_q^{\text{SARMAX}}(k, l) = \frac{|\hat{X}_q^{\text{SIRMAX}}(k, l)|}{\sum_{q=1}^{Q} |\hat{X}_q^{\text{SIRMAX}}(k, l)|} Y(k, l). \tag{5.21}$$

SARMAX provides a lossless decomposition and is commonly used in NMF algorithms. In summary, CMFWISA (King 2012) brings some advantages compared to CMF (Kameoka *et al.* 2009) including fewer parameters, lower computational time, and the additivity for all components within each source in the mixture.

5.4.2 Phase Importance for Signal Reconstruction

We consider the case where an estimate of the spectral magnitude is given, while the spectral phase for reconstruction of the individual sources from the mixture is to be estimated. Table 5.1 lists methods proposed for phase estimation for signal reconstruction in SCSS.

5.4.2.1 Multiple Input Spectrogram Inversion

Gunawan and Sen (2010) proposed an iterative phase estimation approach to synthesize the separated sources from a mixture. The central idea is called *multiple input spectrogram inversion* (MISI). Let $y(n)$ be an observed mixture composed of Q sources given by $y(n) = \sum_{q=1}^{Q} x_q(n)$. Assuming that an initial estimate of the *STFT magnitude* (STFTM) for each source is provided as $|\hat{X}_q(k,l)|$, the MISI algorithm aims to obtain an estimate of the time domain source $\hat{x}_q(n)$ from the given STFTMs and the observed mixture $y(n)$. Figure 5.9 shows a block diagram to illustrate the MISI algorithm.

Let $X_q(k,l)$ be the short-time Fourier transform of the qth source. According to the discussion in Chapter 3, given the magnitude spectrum constraint, an iterative process like the Griffin and Lim update rule (Griffin and Lim 1984) could be used to estimate signals by recovering the missing phase information:

$$\hat{x}_q^{(i+1)}(n) = \frac{\frac{1}{N_{DFT}} \sum_{l=-\infty}^{\infty} w(n-lS) \sum_{k=0}^{N_{DFT}-1} \hat{X}_q^{(i)}(k,l) e^{j2\pi kn/N_{DFT}}}{\sum_{l=-\infty}^{\infty} w^2(n-lS)},$$

$$\hat{X}_q^{(i)}(k,l) = |X_q(k,l)| \frac{\overline{X}_q^{(i)}(k,l)}{|\overline{X}_q^{(i)}(k,l)|}, \tag{5.22}$$

Table 5.1 Phase estimation methods proposed for signal reconstruction in SCSS.

Phase Estimator	Reference
Multiple input spectrogram inversion (MISI)	Gunawan and Sen (2010)
Geometry-based phase estimation	Mowlaee *et al.* (2012a)
Partial phase reconstruction (PPR)	Sturmel and Daudet (2012)
Phase-based informed source separation (PBISS)	Sturmel *et al.* (2012)
Consistent Wiener filter (CWF)	Le Roux *et al.* (2010), Le Roux and Vincent (2013)
Informed source separation using iterative reconstruction (ISSIR)	Sturmel and Daudet (2013)
Sinusoidal-based partial phase reconstruction (SBPPR)	Watanabe and Mowlaee (2013)
Temporal smoothing of unwrapped phase (TSUP)	Mayer and Mowlaee (2015)
Phase reconstruction using linear unwrapping	Magron *et al.* (2015a,b)

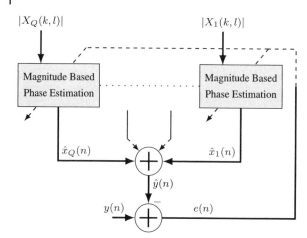

Figure 5.9 Schematic representation of the MISI algorithm (Gunawan and Sen 2010). The spectral magnitudes are combined with the estimated phase to produce time domain signal estimates $\hat{x}_q$. These source estimates are then subtracted from the observed mixture $y(n)$ to produce the remixing error $e(n)$ which is used to refine the phase estimates in iterations.

where S denotes the frame shift and N_{DFT} is the DFT size and $\overline{X}_q^{(i)}(k,l)$ is obtained from STFT of

$$\overline{x}_q^{(i)}(n) = \hat{x}_q^{(i)}(n) + \frac{e^{(i)}(n)}{Q}. \tag{5.23}$$

The remixing error $e^{(i)}(n)$ is calculated by subtracting the superposition of the estimated sources $\hat{x}_q^{(i)}(n)$ from the mixture $y(n)$:

$$e^{(i)}(n) = y(n) - \sum_{q=1}^{Q} \hat{x}_q^{(i)}(n). \tag{5.24}$$

The implication of the source magnitude constraint contributes in shaping the remixing error in that the phase gets re-estimated within the frequency regions with large errors while for regions with small errors the phase is untouched. The iterative procedure is initialized with $e^{(0)}(n) = 0$ and $\hat{x}^{(0)}(n) = y(n)$. As the stopping criterion, Gunawan and Sen proposed to either use a maximum number of iterations $I = 10$ or a minimum error level, such as $e_{\min} = 0.01$. The computational complexity of the MISI approach is $I(2Q + 1)(LN \log_2 N) + 3N$ multiplications, for processing L frames, Q sources, and I maximum iterations.

The MISI approach led to perfect phase reconstruction given a highly accurate source spectral magnitude. The MISI approach highlights the benefit of a closed-loop synthesis algorithm in model-based separation frameworks. The underlying phase estimation routine used with MISI results in significant gains in minimizing the synthesis errors with respect to the time domain mixture for sufficiently accurate given the magnitude spectra of the underlying sources. As one disadvantage, Sturmel and Daudet (2012, 2013) reported that the MISI approach, when used for quantized source magnitude spectra, distributes the remixing error to the silent regions which yields degradation in the source separation performance.

5.4.2.2 Partial Phase Reconstruction
Sturmel and Daudet (2012) proposed the *partial phase reconstruction* (PPR) to synthesize time domain signals given the Wiener filtered spectrogram. The central idea is to apply a threshold on the Wiener gain function to divide it into two regions: a *confidence*

domain, where the mixture phase remains untouched, and its complement, where the GLA update rule is applied.

We consider the scenario where the energy contribution of each source is known while its phase is unknown. The Wiener filter for the qth source is defined as:

$$G_q(k,l) = \frac{|X_q(k,l)|^2}{\sum_{q=1}^{Q} |X_q(k,l)|^2},$$
(5.25)

where the estimated STFT of the qth source is given by

$$\hat{X}_q(k,l) = G_q(k,l)Y(k,l).$$
(5.26)

Sturmel and Daudet argued that the Wiener gain function contains information about signal dominance. In particular, for a Wiener gain $G_q(k,l)$ close to 1, the mixture phase is likely to be dominated by the qth source. They defined a confidence domain based on thresholding the Wiener gain function,

$$\Omega_q = \{(k,l)|G_q(k,l) > \tau_p\},$$
(5.27)

where Ω_q is the confidence domain for the qth source, and τ_p is a fixed threshold. The STFT bins lying in Ω_q are those with higher spectral energy in the qth source.

The confidence domain Ω_q is used to constrain the STFT phase and to enforce the STFT within the Griffin–Lim update rule. For the ith iteration, we get:

$$\hat{X}_q^{(i+1)}(k,l) = \begin{cases} G_q(k,l)|Y(k,l)|e^{j\angle\tilde{X}_q^{(i)}} & \text{for} \quad (k,l) \notin \Omega_q, \\ G_q(k,l)Y(k,l) & \text{for} \quad (k,l) \in \Omega_q, \end{cases}$$
(5.28)

with

$$\tilde{X}_q^{(i)} = \mathcal{G}(\hat{X}_q^{(i)}) = \text{STFT}(\text{iSTFT}(\hat{X}_q^{(i)})),$$
(5.29)

where $\mathcal{G}$ is the consistency operator used as the update rule in GLA.[1] A small maximum number of iterations I was reported to be sufficient as the method initialized with the Wiener filter estimate of the sources converges within a few iterations. For large values of τ_p we get close to the classical Griffin–Lim iterative signal reconstruction. On the other hand, for small values of τ_p, we are close to the original Wiener filter. The two key parameters in this algorithm are the maximum number of iterations I and the threshold τ_p. Through experiments, the authors proposed $\tau_p = 0.65$ and $I = 8$ iterations, which provided a good trade-off between maximum SDR and SIR, both commonly used source separation evaluation measures (Vincent *et al.* 2006).

In the oracle source energy scenario, Sturmel and Daudet showed a net improvement in SIR achieved by PPR, compared to the CWF and the Wiener filter solution. The improvement is achieved with a lower computational complexity with respect to CWF. For a similar computation time, PPR outperforms CWF in terms of SIR, and in terms of SDR only for small computation time below 0.3 seconds (hence, is more suitable for real-time implementation). In contrast to GLA, fewer artifacts were introduced.

5.4.2.3 Informed Source Separation Using Iterative Reconstruction (ISSIR)

In the conventional MISI approach, the remixing error is equally distributed on the underlying sources in the time domain. As some error is distributed into regions where sources are zero, some crosstalk is inevitable. Therefore, Sturmel and Daudet (2013)

---

1 For further details, see Chapter 3.

proposed an activity-based error distribution where the remixing error is distributed where it is needed. They defined the activity domain $\varpi_q(k,l)$ given by:

$$\varpi_q(k,l) = \begin{cases} 1 & G_q(k,l) > \rho_{\text{th}}, \\ 0 & \text{otherwise}, \end{cases} \tag{5.30}$$

with ρ_{th} defined as a constant threshold. The remixing error spreading is performed as follows:

$$\hat{X}_q^{(i+1)}(k,l) = \varpi_q(k,l) \left(\tilde{X}_q^{(i)}(k,l) + \frac{E^{(i)}(k,l)}{\sum_{q'=1}^{Q} \varpi_{q'}(k,l)} \right),$$

$$\tilde{\mathbf{X}}^{(i)} = \mathcal{G}(\hat{\mathbf{X}}_q^{(i)}), \tag{5.31}$$

where $\sum_{q=1}^{Q} \varpi_q(k,l)$ accounts for the overal contributions of sources in time frequency error distribution and the remixing error $E^{(i)}(k,l)$ is defined as

$$E^{(i)}(k,l) = Y(k,l) - \sum_{q=1}^{Q} \hat{X}_q(k,l). \tag{5.32}$$

5.4.2.4 Sinusoidal-Based PPR

The confidence domain proposed in PPR (5.27) results from a fixed threshold and is not signal-adaptive. Therefore, it is sensitive to the dynamic changes in the level of the underlying sources. Watanabe and Mowlaee (2013) proposed an iterative *sinusoidal-based partial phase reconstruction* (SBPPR), introducing two core novelties:

- A sinusoidal model is fitted to the estimated source magnitude spectrum to form a signal-dependent confidence domain for partial phase reconstruction.
- The MISI framework and the distribution of the remixing error at each iteration is extended to sinusoidal signal components.

The update rule for the ith iteration spectrogram for the qth source is given by:

$$\hat{X}_q^{(i+1)}(k,l) = \begin{cases} \tilde{X}_q^{(i)}(k,l) + p(\Omega_{q,\sin}|Y(k,l))E(k,l) & \text{for} \quad (k,l) \notin \Omega_{q,\sin}, \\ G_q(k,l)Y(k,l) + p(\Omega_{q,\sin}|Y(k,l))E(k,l) & \text{for} \quad (k,l) \in \Omega_{q,\sin}, \end{cases} \tag{5.33}$$

with $p(\Omega_{q,\sin}|Y)$ defined as the *a posteriori* probability of selecting $\Omega_{q,\sin}$ for the given mixture as:

$$p(\Omega_{q,\sin}|Y(k,l)) = \frac{p_q p(\Omega_{q,\sin})}{\sum_q p_q}, \tag{5.34}$$

where p_q is the speech presence probability of each sinusoidal component in the qth source.

Figure 5.10 shows an example to illustrate the concept proposed in PPR (Sturmel and Daudet 2012) using a fixed-threshold confidence domain versus the sinusoidal-based confidence domain used in SBPPR (Watanabe and Mowlaee 2013). We consider a mixture of two speech sources mixed at 3 dB. The sinusoid amplitudes (shown in the middle panel) are used to estimate the speech presence probability contribution shown for each source in the bottom panel.

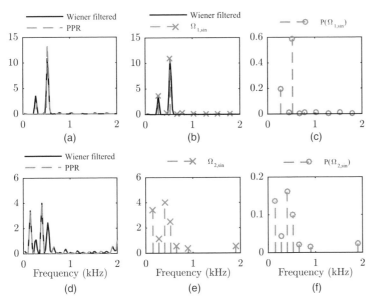

Figure 5.10 Ideal Wiener filter along with the estimated magnitude spectrum using the confidence domain in PPR (Sturmel and Daudet 2012) using fixed threshold $\tau_p = 0.8$ (left). Sinusoidal confidence domain and the estimated magnitude spectrum (middle). Speech presence probability of the sinusoidal confidence domain (right). Results are shown for (top) first speaker, (bottom) second speaker.

5.4.2.5 Spectrogram Consistency

The conventional design of the Wiener time–frequency mask does not consider the spectrogram consistency, i.e., it should correspond to the STFT of a realistic time domain signal. Le Roux *et al.* (2008) proposed to employ this additional constraint of spectrogram consistency into the concept of Wiener filter time–frequency masking which resulted in two methods: one in the time domain and a second one by incorporating a penalty function as the auxiliary constraint.

If we model $X_q(k, l)$ as statistically independent Gaussian random variables with variance $\sigma_q^2(k, l)$, the Wiener filter estimate of $X_1(k, l)$ in the case of $Q = 2$ sources is given by:

$$\hat{X}_1(k, l) = \frac{\sigma_1^2(k, l)}{\sigma_1^2(k, l) + \sigma_2^2(k, l)} Y(k, l), \tag{5.35}$$

as the solution for the minimization of the maximum likelihood objective function as a function of $\mathbf{X} = \mathbf{X}_1$:

$$C(\mathbf{X}) = \sum_{k,l} \alpha(k, l) |X(k, l) - \hat{X}(k, l)|^2, \tag{5.36}$$

where $\alpha(k, l) = \frac{1}{\sigma_1^2(k,l)} + \frac{1}{\sigma_2^2(k,l)}$. With no further constraint, $X(k, l) = \hat{X}(k, l)$ minimizes the objective function. Le Roux *et al.* (2008) introduced the constraint of inconsistency into the objective function of the Wiener filter. The set of consistent spectrograms is the kernel of the consistency operator $\mathcal{G}$ given in (5.29) (Le Roux *et al.* 2008). Now, by taking the consistency definition into the general objective function for the Wiener filter into account, we obtain the following objective function:

$$\tilde{C}(\hat{X}(k,l)) = \sum_{k,l} \alpha(k,l)|\mathring{X}(k,l)|^2, \tag{5.37}$$

where $\mathring{X} = I(\mathbf{X})$. The new objective function given in (5.37) can be optimized in two ways:

- in the time domain by minimizing with respect to the time domain signal $x(n)$
- by considering a penalty function where the weight of the penalty function is chosen large enough or increased during optimization.

The time domain approach applies the consistency as a hard constraint, i.e., $I(\mathbf{X}) = \mathbf{X} - G(\mathbf{X}) = \mathbf{0}$, and requires a large amount of memory space, which results in computationally expensive matrix inversions. Therefore, in the following, we will only consider the penalty-based approach. Considering the l_2 norm of $I(\mathbf{X})$ and including it as a soft penalty constraint, an iterative optimization approach was formulated. Let i be the iteration index and the ith iteration spectrogram as $\hat{\mathbf{X}}^{(i)}$ that is adapted first to $\tilde{\mathbf{X}}^{(i)} = G(\hat{\mathbf{X}}^{(i)})$. Eventually, we obtain the $(i+1)$th spectrogram estimate $\hat{\mathbf{X}}^{(i+1)}$ using an iterative update equation given by:

$$\hat{X}^{(i+1)}(k,l) = \frac{\alpha(k,l)\hat{X}(k,l) + \gamma_c \tilde{X}^{(i)}(k,l)}{\alpha(k,l) + \gamma_c}, \tag{5.38}$$

with γ_c defined asthe key adjustment parameter in CWF. Under the oracle condition of a known power spectrogram of the underlying sources, the method results in significant improvement in SDR (Le Roux *et al.* 2010). They showed that the result of the penalty-based approach approximated the outcome of the time domain method. Employing the iterative procedure for noise reduction, the consistent Wiener filter (CWF) outperforms the conventional Wiener filter in terms of SDR. These results, however, refer to a scenario where the noise PSD is known and the speech spectrum is estimated via spectral subtraction. It was claimed that the residual noise present in the conventional Wiener filter (musical noise) is reduced by the CWF solution.

As shown in Sturmel and Daudet (2012), CWF achieves a lower SIR but an improved SDR compared to PPR. The consistent Wiener filter solution requires an *ad hoc* adjustment of the key parameter γ_c, and shows a slower convergence rate compared to PPR. The soft penalty-based approach achieves the best SDR performance compared to PPR, ISSIR, and the conventional Wiener filter.

Figure 5.11 shows a proof-of-concept result for CWF with known noise variance applied on a -10 dB SNR mixture of speech and cafeteria noise. A PESQ improvement of 1.5 and STOI improvement of 40% is achieved.

5.4.2.6 Geometry-Based Phase Estimation

As discussed in Chapter 3, the minimum reconstruction error criterion together with the geometry of the single-channel speech enhancement problem leads to two sets of phase candidates that are ambiguous in terms of the sign of their phase difference. Similarly, for SCSS, in the case of two sources we have

$$\hat{Y}(k,l) = |\hat{X}_1(k,l)|e^{j\hat{\phi}_{x_1}(k,l)} + |\hat{X}_2(k,l)|e^{j\hat{\phi}_{x_2}(k,l)}, \tag{5.39}$$

$$e(\hat{\phi}_1(k,l), \hat{\phi}_2(k,l)) = |Y(k,l) - \hat{Y}(k,l)|, \tag{5.40}$$

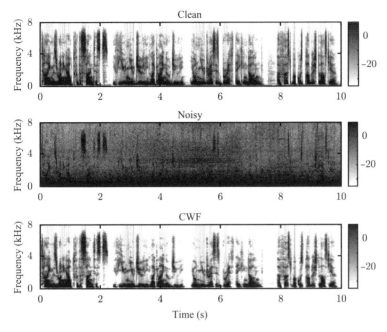

Figure 5.11 Proof-of-concept result for consistent Wiener filter (Le Roux and Vincent 2013) applied on a noisy speech utterance at −10 dB in street noise with a known noise power spectrum. Spectrograms shown in dB for (top) clean, (middle) noisy, and (bottom) CWF outcome.

whereby the best ambiguous pairs denoted by $\{\hat{\phi}_1^{(a)}(k,l), \hat{\phi}_2^{(a)}(k,l)\}$ are given by searching for the minimum reconstruction error e and we have

$$\{\hat{\phi}_1^{(a)}(k,l), \hat{\phi}_2^{(a)}(k,l)\} = \underset{\hat{\phi}_1(k,l),\hat{\phi}_2(k,l)}{\arg\min}\ e(\hat{\phi}_1(k,l), \hat{\phi}_2(k,l)). \tag{5.41}$$

The ambiguous phase candidates only differ in terms of their resulting sign of $\Delta\phi(k,l)$ (or equivalently $\sin\Delta\phi(k,l)$). This ambiguity exists for each source, at each time–frequency bin, and hence is a crux problem to solve. Mowlaee *et al.* (2012a) employed the property of group delay deviation discussed in Chapter 2, assuming that group delay deviation is minimal at spectral peaks of the magnitude spectrum. They further proposed to apply the geometry-based method at harmonics only, where $\{k_h\}_{h=1}^{H}$ denotes the frequency bin at harmonic h with H as the model order of harmonics. For each qth source, the ambiguous candidate phase $\hat{\phi}_1^{(a)}$ is used to set up the distortion constraint criterion d_{GDD} given by:

$$d_{\mathrm{GDD}}(k_h,l) = 1 - \cos\left(\frac{2\pi}{N_{\mathrm{w}}}\tau_{\mathrm{w}} - (\hat{\phi}_1(k_h,l) - \hat{\phi}_1(k_h+1,l))\right), \tag{5.42}$$

where $\tau_{\mathrm{w}} = \frac{N_{\mathrm{w}}-1}{2}$ is denoted as the group delay of a causal window of length N_{w}. Finally, the ambiguity-removed phase estimate $\hat{\phi}_1^*(k_h,l)$ is given by

$$\hat{\phi}_1^*(k_h,l) = \underset{\hat{\phi}_1^{(a)}(k_h,l)}{\arg\min}\ d_{\mathrm{GDD}}(k_h,l). \tag{5.43}$$

The enhanced phase and the spectral magnitudes of the sources are used for signal reconstruction,

$$\hat{x}_1(n) = \text{iSTFT}(|X_1(k,l)|e^{j\phi_1^*(k,l)}).\tag{5.44}$$

A similar procedure is taken for the second source phase spectrum $\phi_2^*(k,l)$ and reconstruction of time signal $\hat{x}_2(n)$.

Throughout their experiments, Mowlaee *et al.* (2012b) demonstrated that for an oracle magnitude spectrum, the geometry-based method results in significant improvement compared to the ideal binary mask (which employs the mixture phase at signal reconstruction) (Wang 2005) and MISI (Gunawan and Sen 2010) approaches in terms of PESQ, segmental SNR, and the mean squared signal reconstruction error. They also considered a semi-blind scenario where the source magnitude spectra were erroneously modeled by additive complex white Gaussian noise. The results showed that for high enough signal-to-quantization error ratio, the geometry-based method outperforms others in terms of signal reconstruction accuracy as well as perceived speech quality.

5.4.2.7 Phase Decomposition and Temporal Smoothing

As discussed in Chapter 3, *temporal smoothing of the unwrapped phase* (TSUP) obtained from a noisy speech observation contributes to improved speech enhancement when used for signal reconstruction. A similar idea was applied by Mayer and Mowlaee (2015) for the SCSS task. The main difficulty in SCSS is the problem of estimating the pitch of multiple sources simultaneously. This problem is referred to as the multi-pitch estimation problem, and has not yet been solved satisfactorily.

Mayer and Mowlaee proposed an improved multi-pitch tracker relying on pre-separated spectral magnitude and an initial pitch estimate. Let $f_{0,q}(l)$ denote the fundamental frequency of the qth source at the lth frame with a minimum and maximum range $[f_{0,\min}, f_{0,\max}]$ used to find an initial $f_{0,q}(l)$ estimate obtained by a *pitch estimation filter with amplitude compression* (PEFAC; Gonzalez and Brookes 2014), known for robust fundamental frequency estimation from noisy speech. $f_{0,q}$ is modeled by a Gaussian distribution with mean and standard deviation used to fit $f_{0,q}$ within the range $[f_{0,\min}, f_{0,\max}]$. To avoid the possibility of a negative frequency, a truncated normal distribution is used. The outcome for the f_0 range before and after pitch refinement is shown in Figure 5.12, concluding that f_0 outliers are significantly reduced. The new obtained frequency range is then reapplied as new input to the pitch estimator of the pre-separated signal. This yields an improved fundamental frequency estimate $\hat{f}_{0,q}$.

The fundamental frequency estimate is used to approximate the linear phase of harmonic trajectories h given by

$$\hat{\psi}_{q,\text{lin}}(h,l) = \frac{2\pi}{f_s} \sum_{l'=0}^{l} h\hat{f}_{0,q}(l')(t(l') - t(l'-1)),\tag{5.45}$$

where $t(l') = t(l'-1) + \frac{1}{4f_0(l-1)}$ are the time instants at the center of the frames and f_s is the sampling frequency. An unwrapped harmonic phase of the source q is then given by subtracting the estimated linear phase from the harmonic phase, and we obtain:

$$\Psi_q(h,l) = \psi_q(h,l) - \hat{\psi}_{q,\text{lin}}(h,l).\tag{5.46}$$

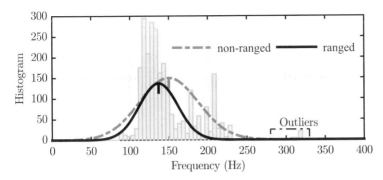

Figure 5.12 Comparison of a ranged (neglecting outliers) and non-ranged (including outliers) frequency distribution. The bars illustrate the histogram of the pre-separated input.

Finally, a temporal smoothing filter is applied to the unwrapped phase by taking the circular mean of $\Psi_q(h, l)$, and for the qth source phase the smoothed unwrapped phase estimate is given by:

$$\hat{\Psi}_q(h, l) = \angle \frac{1}{|\mathcal{R}|} \sum_{l' \in \mathcal{R}} e^{j\Psi_q(h, l')}, \tag{5.47}$$

where $\mathcal{R}$ is the set of neighboring frames lying within a range of 20 ms around frame l. The reconstruction phase is obtained by $\hat{\psi}_q(h, l) = \hat{\Psi}_q(h, l) + \hat{\psi}_{q, \text{lin}}(h, l)$.

The separated spectral magnitudes obtained from IBM and NMF were combined with the estimated phase in order to quantify the achievable improvement in source separation when the mixture phase is replaced with the estimated clean phase spectrum. The temporal smoothing method consistently improves the speech intelligibility of both target and masker at all SNRs compared to the conventional IBM. This is a notable finding as the ideal binary mask was assumed to obtain the highest speech intelligibility performance in SCSS (Wang and Brown 2006). Furthermore, the phase enhancement contributes to improve the perceived quality of the IBM, which is known to suffer from low perceptual quality (Williamson *et al.* 2014). The phase enhancement combined with estimated NMF-separated spectral magnitudes achieves consistent improvement in PESQ, STOI, SIR, and SDR scores. A large improvement from the use of clean phase was reported. All these findings emphasize the importance of a proper phase modification to push the limited achievable performance of IBM or NMF, using mixture phase. Experiment 5.1 in Section 5.5.1 presents a proof of concept to demonstrate the outcome of this approach.

5.4.2.8 Phase Reconstruction of Spectrograms with Linear Unwrapping

Magron *et al.* (2015a,b) studied the phase recovery in a NMF framework for audio source separation. They considered temporal and spectral coherence, where temporal coherence refers to horizontal unwrapping over time frames while the spectral coherence refers to vertical unwrapping along frequency bins.

For horizontal unwrapping (along time), we consider a localized complex-valued sinusoid $x(n) = A e^{j(2\pi f_0 n / f_s + \phi_0)}$. After applying a window function $w(n)$, the STFT of a sinusoid is given by

$$X(k,l) = Ae^{j(2\pi \frac{l}{f_s} Sl + \phi_0)} We^{j2\pi \left(\frac{k}{N_{DFT}} - \frac{f_0}{f_s} \right)}, \tag{5.48}$$

where $W(e^{j\omega})$ is the discrete-time Fourier transform of the analysis window $w(n)$, S is the frameshift, and N_{DFT} the Fourier transform length. For the unwrapped phase of successive frames in time we have

$$\hat{\phi}(k,l) = \phi(k,l-1) + 2\pi Sf_0(l/f_s). \tag{5.49}$$

This relation can be used to predict the phase across time if the frequency variation is small between two successive frames. The generalization of this concept to all frames requires accurate knowledge about instantaneous frequency.

To cope with the onsets, for unwrapping along frequency channels, Magron *et al.* (2015b) considered an impulse signal model $x(n) = A\delta(n - n_0)$ with the attack time n_0 and $A > 0$ as its amplitude, yielding the STFT representation

$$X(k,l) = Aw(n_0 - lS)e^{-j2\pi \frac{k}{N_{DFT}}(n_0 - lS)}, \tag{5.50}$$

with the DFT length N_{DFT} and the window function $w(n)$. From (5.50), we get:

$$\phi(k,l) = \phi(k-1,l) - \frac{2\pi}{N_{DFT}}(n_0(k) - lS), \tag{5.51}$$

where $n_0(k)$ refers to the *instantaneous attack time* in each frequency channel and $\phi(0,l) = 0$. Equations (5.49) and (5.51) are similar in that an impulse in frequency corresponds to a sinusoid in time. The vertical unwrapping in (5.51) requires an attack time estimation. For this purpose, Magron *et al.* (2015b) used either a least-squares method or a temporal quadratic interpolation FFT (Abe and Smith 2004) and updated the phase with (5.51). Experiments were conducted to evaluate the performance of the phase retrieval methods in SCSS applications. A comparison with the GLA showed improved SDR.

5.4.3 Phase-Aware Time–Frequency Masks

5.4.3.1 Phase-Insensitive Masks

In order to obtain an estimate of the desired source target, time–frequency filtering methods are quite common to apply to the mixture. Given an estimated time–frequency mask $\hat{G}(k,l)$, the target speech is estimated by $\hat{X}_1(k,l) = \hat{G}(k,l)Y(k,l)$. The objective functions used for the derivation of a desired time–frequency mask G^* are divided into two groups, namely:

- *signal approximation* (SA), which considers an objective measure $D_{SA}(\hat{G}) = D(X\|\hat{X})$ defined between the filtered signal $\hat{X} = \hat{G} \otimes Y$, where the bold symbols refer to the matrix representations;
- *mask approximation* (MA), which computes a target mask $G^*(X, Y)$, and measures the error between the estimated mask $\hat{G}$ and the target mask G^* using an objective measure $D_{MA}(\hat{G}) = D(\hat{G}\|G^*)$. The most commonly used mask functions are the *ideal binary mask* (IBM; Li and Wang 2009) and the *ideal ratio mask* (IRM; Narayanan and Wang 2013).

Table 5.2 lists various masking functions used for source separation, along with their formulas.

Table 5.2 List of time–frequency masks used for SCSS, considering two sources.

Mask Function	Formula
Ideal binary mask (IBM)	$G_{\text{IBM}(k,l)} = \delta(|X_1(k,l)| > |X_2(k,l)|)$
Ideal ratio mask (IRM)	$G_{\text{IRM}(k,l)} = \left(\dfrac{|X_1(k,l)|^2}{|X_1(k,l)|^2 + |X_2(k,l)|^2} \right)^{1/2}$
Wiener filter	$G(k,l) = \dfrac{|X_1(k,l)|^2}{|X_1(k,l)|^2 + |X_2(k,l)|^2}$
Phase-sensitive filter (PSF)	$G_{\text{PSF}(k,l)} = \dfrac{|X_1(k,l)|}{|Y(k,l)|} \cos(\phi_y(k,l) - \phi_1(k,l))$
Complex mask (CM)	$G_{\text{CM}(k,l)} = \dfrac{|X_1(k,l)|}{|Y(k,l)|} e^{j(\phi_1(k,l) - \phi_y(k,l))}$
Complex ratio mask (CRM)	$G_{\text{CRM}(k,l)} = G_{\Re,\text{CRM}(k,l)} + jG_{\Im,\text{CRM}(k,l)}$

5.4.3.2 Phase-Sensitive Mask

Assuming a real-valued mask G, the phase-sensitive mask function is given by minimizing $|\varepsilon|^2$ with $\varepsilon = G(k,l)|Y(k,l)|e^{j\phi_y(k,l)} - |X(k,l)|e^{j\phi_x(k,l)}$, and is given by:

$$\hat{G}_{\text{PSF}}(k,l) = \arg\min_{G(k,l)} \quad (G(k,l)|Y(k,l)|e^{j\phi_y(k,l)} - |X(k,l)|e^{j\phi_x(k,l)})^2. \tag{5.52}$$

By setting the derivative of $|\varepsilon|^2$ with respect to G equal to zero we obtain the *phase-sensitive filter* (PSF):

$$G_{\text{PSF}}(k,l) = \frac{|X(k,l)|}{|Y(k,l)|} \cos(\phi_y(k,l) - \phi_x(k,l)). \tag{5.53}$$

Erdogan *et al.* (2015) first proposed the PSF as an objective function, optimized for maximizing the SNR of the reconstructed signals. Assuming a known phase spectrum, the phase-sensitive filter significantly outperforms other filters in terms of SDR. This improved performance can be explained by the shrinking amount of noise suppression, thanks to the cosine of the phase deviation term. Later, Weninger *et al.* (2015) reported that the word error rate obtained by a neural network[2] ASR back end trained on the PSF is highly correlated with the SDR scores obtained at the front end.

5.4.3.3 Complex Ratio Mask

Williamson *et al.* (2016) proposed an ideal *complex ratio mask* (CRM) and its estimation using a deep neural network. The desired *complex mask* (CM) G_{CRM} can retrieve the clean speech spectrum:

$$X_1(k,l) = G_{\text{CRM}}(k,l)Y(k,l). \tag{5.54}$$

Representing the signals by their real and imaginary parts $X_1(k,l) = X_{1,\Re}(k,l) + jX_{1,\Im}(k,l)$, $Y(k,l) = Y_{\Re}(k,l) + jY_{\Im}(k,l)$, as well as the mask $G_{\text{CRM}}(k,l) = G_{\Re,\text{CRM}}(k,l) + jG_{\Im,\text{CRM}}(k,l)$, and rewriting (5.54), the clean signal is given by:

$$X_{1,\Re}(k,l) = G_{\Re,\text{CRM}}(k,l)Y_{\Re}(k,l) - G_{\Im,\text{CRM}}(k,l)Y_{\Im}(k,l),$$
$$X_{1,\Im}(k,l) = G_{\Re,\text{CRM}}(k,l)Y_{\Im}(k,l) + G_{\Im,\text{CRM}}(k,l)Y_{\Re}(k,l). \tag{5.55}$$

Solving for the real and imaginary parts of the desired mask, we get the CRM:

$$G_{\text{CRM}}(k,l) = G_{\Re,\text{CRM}}(k,l) + jG_{\Im,\text{CRM}}(k,l), \tag{5.56}$$

---

2 A long short-term memory recurrent neural network (LSTM-RNN) was used.

where its real and imaginary parts are

$$G_{\Re,\text{CRM}}(k,l) = \frac{Y_{\Re}(k,l)X_{1,\Re}(k,l) + Y_{\Im}(k,l)X_{1,\Im}(k,l)}{Y_{\Re}^2(k,l) + Y_{\Im}^2(k,l)}, \tag{5.57}$$

$$G_{\Im,\text{CRM}}(k,l) = \frac{Y_{\Re}(k,l)X_{1,\Im}(k,l) - Y_{\Im}(k,l)X_{1,\Re}(k,l)}{Y_{\Re}^2(k,l) + Y_{\Im}^2(k,l)}. \tag{5.58}$$

5.4.3.4 Complex Mask

In contrast to the PSF, if we assume that G_{CM} is a complex value,

$$G_{\text{CM}}(k,l) = |G_{\text{CM}}(k,l)|e^{j\angle G_{\text{CM}}(k,l)}, \tag{5.59}$$

and then solve the minimization problem,

$$G_{\text{CM}}(k,l) = \arg\min_{G_{\text{CM}}(k,l)} \quad (G_{\text{CM}}(k,l)|Y(k,l)|e^{j\phi_y(k,l)} - |X_1(k,l)|e^{j\phi_x(k,l)})^2, \tag{5.60}$$

and taking the derivative of $|\epsilon| = \epsilon\epsilon^*$, defining

$$\epsilon = G_{\text{CM}}(k,l)|Y(k,l)|e^{j\phi_y(k,l)} - |X(k,l)|e^{j\phi_x(k,l)}, \tag{5.61}$$

and solving it for zero, we obtain the complex mask function:

$$G_{\text{CM}}(k,l) = \frac{|X(k,l)|}{|Y(k,l)|}e^{j(\phi_x(k,l)-\phi_y(k,l))}. \tag{5.62}$$

If we apply the complex mask function from (5.62) to the mixture, we obtain:

$$X(k,l) = G(k,l)Y(k,l) = |X(k,l)|e^{j\phi_x(k,l)}, \tag{5.63}$$

which recovers the complex DFT coefficient $X(k,l)$ in terms of both magnitude and phase. It is important to note that it is straightforward to show that the CM, $G_{\text{cm}}(k,l)$ (5.62), and the CRM, $G_{\text{CRM}}(k,l)$ in (5.56), are mathematically equivalent, since the real and imaginary parts of CRM represent the complex exponential using Euler's equation.

5.4.4 Phase Importance in Signal Interaction Models

So far, we have discussed the importance of phase in spectral modification of the matrix factorization framework, time–frequency masking, and reconstructing separate sources. The phase may contribute to estimating the spectral amplitude when used in a signal interaction model. A signal interaction model describes the distribution of a mixed signal given its underlying sources. The goal is to find out how the desired target and masker signals interact with each other to form the observed mixed signal. An interaction model could be used either in a complete feature domain (e.g., complex spectrum) or a lossy feature domain, for example, features extracted for ASR. In the following, we will only consider interaction models in the full spectral domain (for a full overview of other interaction models, see (Hershey *et al.* 2013, Chapter 12)). Table 5.3 lists the signal interaction functions.

The first and the simplest signal interaction model is called *mixture maximization* (MIXMAX), first proposed for automatic speech recognition (Varga and Moore 1990; Nadas *et al.* 1989). The core idea is to approximate the mixture by the element-wise individual maximum of the two sources:

$$|\hat{Y}(k,l)| \approx \exp\left(\max(\log|X_1(k,l)|, \log|X_2(k,l)|)\right). \tag{5.64}$$

Table 5.3 List of signal interaction functions.

Signal Interaction Function	Estimated Mixture $\hat{Y}(k, l)$
MIXMAX	$\max(\log|X_1(k, l)|, \log|X_2(k, l)|)$
Log-power MMSE	$\sqrt{|X_1(k, l)|^2 + |X_2(k, l)|^2}$
Linear interaction	$|X_1(k, l)| + |X_2(k, l)|$
Amplitude MMSE	$\frac{|X_1(k,l)| + |X_2(k,l)|}{\pi} \int_0^\pi \sqrt{1 - \eta^2 \sin^2\psi}\, d\psi$
Phase-sensitive	$\sqrt{|X_1(k, l)|^2 + |X_2(k, l)|^2 + 2|X_1(k, l)||X_2(k, l)|\cos(\Delta_s)}$

Radfar *et al.* (2006) showed that the MIXMAX approximation is a non-linear minimum mean square error (MMSE) estimator when the phase difference between the sources follows a uniform distribution. MIXMAX has been successfully used for speaker recognition (Rose *et al.* 1994), single-channel speech enhancement (Burshtein and Gannot 2002), and SCSS (Roweis 2003; Reddy and Raj 2004).

While MIXMAX entirely neglects the phase information, assuming orthogonality of the sources (which yields $\cos(\Delta\phi) = 0$), the power MMSE approximation of $|Y(k, l)|$ is given by (Reddy and Raj 2004):

$$|\hat{Y}(k, l)| = \sqrt{|X_1(k, l)|^2 + |X_2(k, l)|^2}. \tag{5.65}$$

Using the geometry of source separation,

$$|\hat{Y}(k, l)| = \sqrt{|X_1(k, l)|^2 + |X_2(k, l)|^2 + 2|X_1(k, l)||X_2(k, l)|\cos(\Delta\phi(k, l))}, \tag{5.66}$$

the MMSE spectral amplitude estimate under the assumption of a uniform phase difference prior gives the following signal interaction function (Mowlaee *et al.* 2008):

$$|\hat{Y}(k, l)| = \frac{|X_1(k, l)| + |X_2(k, l)|}{\pi} \int_0^\pi \sqrt{1 - \eta^2 \sin^2 t'}\, dt', \tag{5.67}$$

where $\eta = \frac{2\sqrt{|X_1(k,l)||X_2(k,l)|}}{|X_1(k,l)| + |X_2(k,l)|}$. The integral in (5.67),

$$E(\eta, \pi) = \int_0^\pi \sqrt{1 - \eta^2 \sin^2 t'}\, dt', \tag{5.68}$$

is the complete elliptic integral of the second kind (Spiegel 1992). This integral can be approximated by the elliptic series[3] (Spiegel 1992):

$$E(\eta, \pi) = \pi \left[1 - \left(\frac{1}{2}\right)^2 \eta^2 - \left(\frac{1 \times 3}{2 \times 4}\right)^2 \frac{\eta^4}{3} - \left(\frac{1 \times 3 \times 5}{2 \times 4 \times 6}\right)^2 \frac{\eta^6}{5} - \cdots \right]. \tag{5.69}$$

By replacing the integral with its series representation, we obtain:

$$|\hat{Y}(k, l)| = \frac{|X_1(k, l)| + |X_2(k, l)|}{\pi} E(\eta, \pi). \tag{5.70}$$

---

3 Which converges quickly using its first three terms.

The linear interaction model, in contrast to MIXMAX (5.64), works in the magnitude domain (Rose *et al.* 1994). Peharz and Pernkopf (2012) suggested a linear interaction of DFT spectra,

$$|\hat{Y}(k, l)| = |X_1(k, l)| + |X_2(k, l)|, \tag{5.71}$$

and reported better source separation than MIXMAX. Mowlaee and Martin (2012), as a phase-sensitive interaction model, took into account the complete form relying on the phase difference $\Delta\phi(k, l)$ (5.66), by quantizing it using a quantization step of Δ_s, given by

$$|\hat{Y}(k, l)| = \sqrt{|X_1(k, l)|^2 + |X_2(k, l)|^2 + 2|X_1(k, l)||X_2(k, l)|\cos(\Delta_s)}, \tag{5.72}$$

called generalized MMSE. A grid search over Δ_s was performed to find the optimal $\Delta\phi(k, l)$ using a fine enough resolution (e.g., $\Delta_s = \frac{\pi}{100}$; Mowlaee and Martin 2012).

5.5 Experiments

In this section, we present several experiments to demonstrate the importance of phase-aware processing in single-channel source separation. The experiments are:

- Experiment 5.1: Phase estimation for proof-of-concept signal reconstruction
- Experiment 5.2: Comparative study of GLA-based phase reconstruction methods
- Experiment 5.3: Phase-aware time–frequency masks
- Experiment 5.4: Phase-sensitive interaction functions
- Experiment 5.5: Complex matrix factorization.

5.5.1 Experiment 5.1: Phase Estimation for Proof-of-Concept Signal Reconstruction

As our first experiment, we consider a proof of concept to demonstrate the effectiveness of TSUP (explained in Section 5.4.2.7). We consider this phase reconstruction method on top of IRM, which is known to have a very high intelligibility. The mixture file was created at 0 dB SSR using a female speaker uttering "bin blue at d one soon" and a male speaker uttering "bin blue at e one soon," chosen from the GRID corpus (Cooke *et al.* 2006).

Figure 5.13 shows the spectrograms of the target signal, the mixture, IRM, and IRM combined with the TSUP phase estimator. The improvements in the objective metrics are reported at the top of the panels. TSUP, on top of IRM, is capable of improving all objective evaluation criteria. Comparing Figure 5.13(c) and (d), the restoration of the harmonic structure is highlighted, observing that the energy between harmonics is attenuated. Harmonics related to the desired speaker are well identified given a proper fundamental frequency estimate. The MATLAB® implementation for this experiment can be found in *Exp5_1.m*, available in the *PhaseLab Toolbox* (see the appendix).

5.5.2 Experiment 5.2: Comparative Study of GLA-Based Phase Reconstruction Methods

In this experiment, we consider different phase reconstruction methods proposed for SCSS: MISI (Gunawan and Sen 2010), ISSIR (Sturmel and Daudet 2013), PPR (Sturmel and Daudet 2012), SBPPR (Watanabe and Mowlaee 2013), and the consistent Wiener

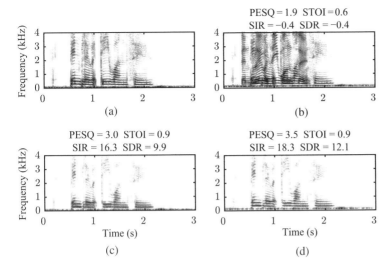

Figure 5.13 Proof of concept showing the outcome of applying temporal smoothing phase estimation on the ideal ratio mask (d). The clean target reference (a), mixture (b), and the ideal ratio mask outcome (c) are shown for comparison.

filter (Le Roux *et al.* 2010). All results are averaged over five utterances selected from GRID mixed at 0 dB SSR. We quantify the source separation performance using the so-called *blind source separation evaluation* (BSS EVAL) measures (Vincent *et al.* 2006), which consist of three SNR-based scores (SDR, SIR, and SAR) that are commonly used to evaluate the separation performance of SCSS methods.

5.5.2.1 Convergence Analysis
In Figure 5.14, we present the convergence behavior of the GLA-based methods (see Table 5.1). We first assume that the source magnitude spectra are known. A maximum number of 200 iterations is considered to ensure that all algorithms converge. ISSIR achieves slightly worse performance compared to PRR, as reported in Sturmel and Daudet (2012). MISI provides large SIR at the expense of relatively low SDR and SAR scores, confirming the observations in Sturmel *et al.* (2012). CWF achieves the best SDR but suffers from slow convergence and a low SIR performance.

5.5.2.2 Quantized Scenario
To demonstrate the performance of phase reconstruction methods in SCSS, we relax the oracle magnitude spectrum scenario and consider quantized source magnitude spectra. The current experiment will highlight the robustness of the phase estimation methods for SCSS when erroneous estimates for source magnitude spectra are provided. To model such erroneous magnitude spectra, we produce a noisy version of the complex signal spectra by adding white Gaussian noise within the range of $[0, 50]$ dB with 10 dB steps, as depicted in Figure 5.15. The quantized source magnitude spectra $X_q(k, l)$ are passed to a Wiener filter, which is applied to the mixture to produce quantized separated signals. Following the recommendation in Sturmel and Daudet (2012), we used a maximum number of iterations $I = 20$ and a fixed threshold of $\tau_p = 0.8$. For ISSIR (Sturmel and Daudet 2013), we chose $\rho_{th} = 0.01$ and $I = 100$, and for MISI, 200 iterations.

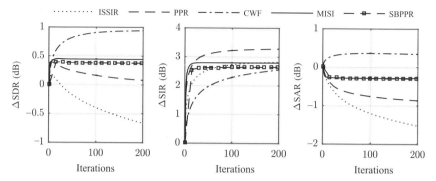

Figure 5.14 Convergence analysis for GL-based methods and their performance reported in terms of ΔSDR, ΔSIR, and ΔSAR compared to Wiener filtering using the mixture phase as baseline (Watanabe and Mowlaee 2013).

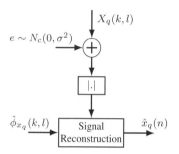

Figure 5.15 Quantized magnitude spectrum inspired by Mowlaee *et al.* (2012a) obtained by adding additive white Gaussian noise *e* to each signal source X_q.

The results are shown in Figure 5.16. Here, we report the results as the relative improvement with respect to the Wiener filter with mixture phase, denoted as delta scores. CWF achieves the best SDR performance at the expense of a lower SIR result. The convergence of the consistency method is worse than the others as it requires at least 50 iterations. MISI provides improved SIR performance at the expense of a degraded SDR outcome. SBPPR and PPR both show a good balance between a large SIR but a slight reduction in SDR. These results emphasize the trade offs that have to be made between the evaluation metrics. The MATLAB® implementation for this experiment can be found in *Exp5_2.m*, provided in the *PhaseLab Toolbox* (for more details, see the appendix).

5.5.3 Experiment 5.3: Phase-Aware Time–Frequency Mask

As a proof of concept, we examine the effectiveness of the *complex ratio mask* for an oracle phase scenario. In this experiment, the ideal ratio mask is used to produce the estimated source magnitude spectra. The results are shown in Figure 5.17 in terms of ΔSDR and ΔSIR versus the input signal-to-noise ratio ranging from −9 dB to 9 dB in 3 dB steps. The outcome of the CRM is also included as a benchmark. The results are averaged over ten TIMIT utterances corrupted with babble noise. The complex mask given in (5.62) provides consistent improvement compared to the ideal ratio mask with the mixture phase in terms of SDR and SIR. The complex ratio mask leads to less improvement in SDR and SIR, while $\cos(\Delta\phi)$ only improves SDR.

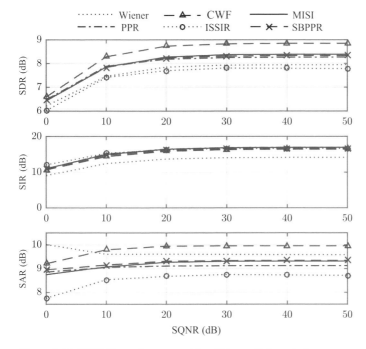

Figure 5.16 BSS EVAL results reported for different GL-based phase reconstruction methods versus different quantization levels: (top) SDR, (middle) SIR, (bottom) SAR (all in dB).

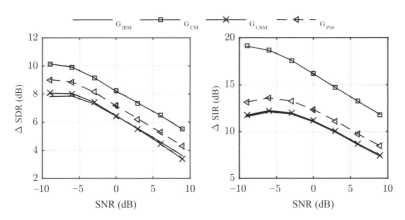

Figure 5.17 ΔSDR (left) and ΔSIR (right) in dB for different masks applied to the mixture spectrum, assuming that the phase spectrum is known.

To further explore the effectiveness of the complex ratio mask, we consider a male speaker saying "They inhabit a secret world centered on go codes and gold phones" corrupted with babble noise at −3 dB. The spectrograms for the clean and noisy signals are shown in Figure 5.18 together with the separated signals obtained from IBM and CM. Thanks to phase-aware processing, certain missing harmonic structure is

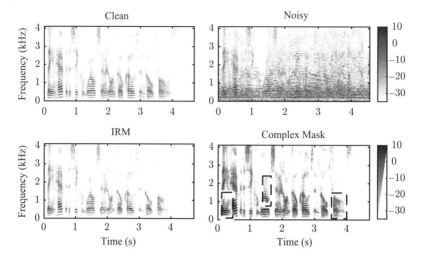

Figure 5.18 Proof-of-concept result obtained by a complex mask applied on noisy male utterance at −3 dB. Shown are clean speech, noisy, IRM, complex mask.

retrieved (marked with dashed-line rectangles). The MATLAB® implementation for this experiment can be found in *Exp5_3.m*, available in the *PhaseLab Toolbox* (see the appendix).

5.5.4 Experiment 5.4: Phase-Sensitive Interaction Functions

To study the importance of phase difference on the achievable accuracy of the signal interaction function, we consider an experiment with $|X_1(k, l)| = 1$ while $|X_2(k, l)|$ is swept according to SSR between −8 dB and 8 dB. The phase difference $\Delta\phi$ is quantized using $\Delta_s \in [-\pi, \pi[$ and various quantization steps of $\Delta_s \in \{\pi/5, \pi/10, \pi/20, \pi/100, \pi/200\}$. The left panel in Figure 5.19 shows a comparison between different mixture approximations versus SSR. Both MIXMAX and log-power MMSE estimators result in a bias compared to the exact value of the mixture. The right panel in Figure 5.19 shows the MSE obtained by different signal interaction functions averaged over five speech signals. The window length was set to 32 ms with 8 ms frameshift at 16 kHz sampling frequency. Again, the elliptic integral provides a more accurate approximation than log and MMSE power estimators, while the phase-sensitive one leads to the more accurate mixture estimate versus different quantization steps.

5.5.5 Experiment 5.5: Complex Matrix Factorization

We consider a proof-of-concept experiment to demonstrate the outcome of applying the CMF algorithm on a mixture. The mixture contains a male target speaker uttering "bin blue at c six now" masked by a female speaker uttering "bin red in d one soon" at 0 dB SSR. The spectrograms of the separation outcomes are shown in Figure 5.20 for clean, mixture, and NMF, CMF-WISA as well as CMF output. For this experiment

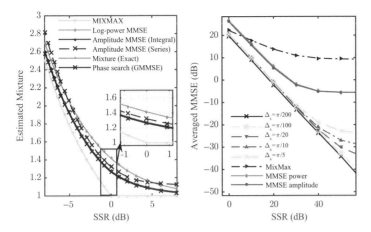

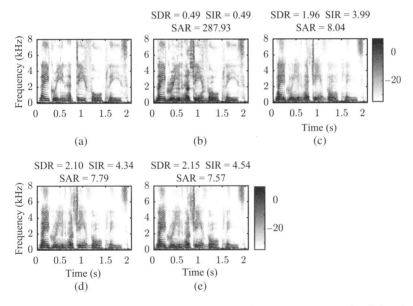

Figure 5.19 The mixture amplitude $|\hat{Y}(k, l)|$ obtained from different signal interaction functions (left). Mean square error averaged over speech frames achieved by different signal interaction functions (right; Mowlaee and Martin 2012).

Figure 5.20 Proof-of-concept result for (a) clean male utterance corrupted with female utterance as masker, (b) mixed signal at 0 dB, (c) NMF outcome, (d) CMF, and (e) CMF-WISA.

we used the *CMF Toolbox*[4] with speaker-dependent dictionaries. The improvements in objective metrics are reported above the panels using the blind source-separation toolbox (Vincent *et al.* 2006). The MATLAB® code for this experiment is found in *Exp5_5.m* provided by the *PhaseLab Toolbox* (see the appendix).

4 Brian King and Les Atlas, *Complex Matrix Factorization Toolbox Version 1.0 for MATLAB*, downloadable at https://sites.google.com/a/uw.edu/isdl/projects/cmf-toolbox, University of Washington, September 2012.

5.6 Summary

This chapter discussed *single-channel source separation* (SCSS) to demonstrate the importance of phase processing in speech communication applications. Starting with a review of existing methods, categorized as source-driven and model-driven approaches, the phase-insensitive methods were first reviewed. Several methods to estimate phase for signal reconstruction in SCSS were discussed, with a comparative study to evaluate their performance in different scenarios.

We provided a review of complex matrix factorization (CMF) and compared it with conventional non-negative matrix factorization (NMF) where additivity of source magnitude spectra is assumed. Through various proof-of-concept experiments, we demonstrated the achievable improvement by the CMF algorithm (phase-aware) versus NMF (phase-unaware). We studied the phase-sensitive signal interaction functions leading to a more accurate approximation of the mixture compared to phase-insensitive ones where phase information is neglected. Several phase-aware versions for time–frequency masks were explained and compared to traditional binary or ratio masks.

References

M. Abe and J. O. Smith, Design criteria for simple sinusoidal parameter estimation based on quadratic interpolation of FFT magnitude peaks, *Audio Engineering Society*, 2004.

S. Araki, F. Nesta, E. Vincent, Z. Koldovský, G. Nolte, A. Ziehe, and A. Benichoux, *The 2011 Signal Separation Evaluation Campaign (SiSEC2011): Audio Source Separation*, Proceedings of the International Conference on Latent Variable Analysis and Signal Separation (LVA/ICA), pp. 414–422, 2012.

J. Barker, E. Vincent, N. Ma, H. Christensen, and P. Green, The PASCAL CHiME speech separation and recognition challenge, *Computer Speech & Language*, vol. 27, no. 3, pp. 621–633, 2013.

J. Barker, R. Marxer, E. Vincent, and S. Watanabe, *The Third CHiME Speech Separation and Recognition Challenge: Dataset, Task and Baselines*, Proceedings of the IEEE International Conference on Automatic Speech Recognition and Understanding Workshop (ASRU 2015), 2015.

A. S. Bregman, *Auditory Scene Analysis*, MIT Press, 1990.

D. Burshtein and S. Gannot, Speech enhancement using a mixture-maximum model, *IEEE Transactions on Speech and Audio Processing*, vol. 10, no. 6, pp. 341–351, 2002.

E. C. Cherry, Some experiments on the recognition of speech, with one and with two ears, *The Journal of the Acoustical Society of America*, vol. 25, pp. 975–979, 1953.

M. Cooke, J. Barker, S. Cunningham, and X. Shao, An audio-visual corpus for speech perception and automatic speech recognition, *The Journal of the Acoustical Society of America*, vol. 120, pp. 2421–2424, 2006.

M. Cooke, J. R. Hershey, and S. J. Rennie, Monaural speech separation and recognition challenge, *Computer Speech & Language*, vol. 24, no. 1, p. 1–15, 2010.

H. Erdogan, J. R. Hershey, S. Watanabe, and J. Le Roux, *Phase-Sensitive and Recognition-Boosted Speech Separation using Deep Recurrent Neural Networks*, Proceedings of the IEEE International Conference on Acoustics, Speech and Signal Processing (ICASSP), 2015.

C. Fevotte and J. Idier. Algorithms for nonnegative matrix factorization with the beta-divergence, *Neural Computation*, vol. 23, no. 9, pp. 2421–2456, 2011.

S. Gonzales and M. Brookes, PEFAC: A pitch estimation algorithm robust to high levels of Noise, *IEEE/ACM Transactions on Audio, Speech, and Language Processing*, vol. 22, no. 2, pp. 518–530, 2014.

D. Griffin and J. Lim, Signal estimation from modified short-time Fourier transform, *IEEE Transactions on Acoustics, Speech, and Signal Processing*, vol. 32, no. 2, pp. 236–243, 1984.

D. Gunawan and D. Sen, Iterative phase estimation for the synthesis of separated sources from single-channel mixtures, *IEEE Signal Processing Letters*, vol. 17, no. 5, pp. 421–424, 2010.

J. R. Hershey, S. J. Rennie, P. A. Olsen, and T. T. Kristjansson, Super-human multi-talker speech recognition: A graphical modeling approach, *Computer Speech & Language*, vol. 24, no. 1, pp. 45–66, 2010.

J. R. Hershey, S. J. Rennie, and J. Le Roux, Factorial models for noise robust speech recognition, in *Techniques for Noise Robustness in Automatic Speech Recognition* (eds T. Virtanen, R. Singh, and B. Raj), Wiley, 2013.

G. Hu and D. L. Wang, Monaural speech segregation based on pitch tracking and amplitude modulation, *IEEE Transactions on Neural Networks*, vol. 15, no. 5, pp. 1135–1150, 2004.

Z. Jin and D. Wang, A supervised learning approach to monaural segregation of reverberant speech, *IEEE Transactions on Audio, Speech, and Language Processing*, vol. 17, no. 4, pp. 625–638, 2009.

H. Kameoka, N. Ono, K. Kashino, and S. Sagayama, *Complex NMF: A New Sparse Representation for Acoustic Signals*, Proceedings of the IEEE International Conference on Acoustics, Speech and Signal Processing (ICASSP), pp. 3437–3440, 2009.

B. J. King, *New Methods of Complex Matrix Factorization for Single-Channel Source Separation and Analysis*, PhD thesis, University of Washington, 2012.

B. J. King and L. Atlas, Single-channel source separation using complex matrix factorization, *IEEE Transactions on Audio, Speech, and Language Processing*, vol. 19, no. 8, pp. 2591–2597, 2011.

J. Le Roux and E. Vincent, Consistent Wiener filtering for audio source separation, *IEEE Signal Processing Letters*, vol. 20, no. 3, pp. 217–220, 2013.

J. Le Roux, N. Ono, and S. Sagayama, *Explicit Consistency Constraints for STFT Spectrograms and their Application to Phase Reconstruction*, Proc. Workshop on Statistical and Perceptual Audition (SAPA), 2008.

J. Le Roux, E. Vincent, Y. Mizuno, H. Kameoka, N. Ono, and S. Sagayama, Consistent Wiener filtering: Generalized time—frequency masking respecting spectrogram consistency, in *Latent Variable Analysis and Signal Separation*, vol. 6365 of Lecture Notes in Computer Science, pp. 89–96, 2010.

Y. Li and D. Wang, *On the Optimality of Ideal Binary Time—Frequency Masks*, Proceedings of the IEEE International Conference on Acoustics, Speech and Signal Processing (ICASSP), pp. 3501–3504, 2008.

Y. Li and D. Wang, On the optimality of ideal binary time–frequency masks, *Speech Communication*, vol. 51, no. 3, 2009.

P. Magron, B. Badeau, and B. David, *Phase Recovery in NMF for Audio Source Separation: An Insightful Benchmark*, Proceedings of the IEEE International Conference on Acoustics, Speech and Signal Processing (ICASSP), pp. 81–85, 2015a.

P. Magron, R. Badeau, and B. David, *Phase Reconstruction of Spectrograms with linear Unwrapping: Application to Audio Signal Restoration*, Proceedings of the European Signal Processing Conference (EUSIPCO), 2015b.

F. Mayer and P. Mowlaee, *Improved Phase Reconstruction in Single-Channel Speech Separation*, Proceedings of the International Conference on Spoken Language Processing (INTERSPEECH), pp. 1795–1799, 2015.

N. Mohammadiha, P. Smaragdis, and A. Leijon, Supervised and unsupervised speech enhancement using nonnegative matrix factorization, *IEEE Transactions on Audio, Speech, and Language Processing*, vol. 21, no. 10, pp. 2140–2151, 2013.

P. Mowlaee, *New Stategies for Single-Channel Speech Separation*, PhD thesis, Institut for Elektroniske Systemer, Aalborg Universitet, 2010.

P. Mowlaee and R. Martin, *On Phase Importance in Parameter Estimation for Single-Channel Source Separation*, Proceedings of the International Workshop on Acoustic Signal Enhancement (IWAENC), pp. 1–4, 2012.

P. Mowlaee, A. Sayadiyan, and M. Sheikhan, *Optimum Mixture Estimator for Single-Channel Speech Separation*, Proceedings of the International Symposium on Telecommunications (IST), pp. 543–547, 2008.

P. Mowlaee, R. Saeidi, and R. Martin, *Phase Estimation for Signal Reconstruction in Single-Channel Speech Separation*, Proceedings of the International Conference on Spoken Language Processing (INTERSPEECH), pp. 1548–1551, 2012a.

P. Mowlaee, R. Saeidi, M. G. Christensen, Z. H. Tan, T. Kinnunen, P. Fränti, and S. H. Jensen, A joint approach for single-channel speaker identification and speech separation, *IEEE Transactions on Audio, Speech, and Language Processing*, vol. 20, no. 9, pp. 2586–2601, 2012b.

A. Nadas, D. Nahamoo, and M. Picheny, Speech recognition using noise-adaptive prototypes, *IEEE Transactions on Speech and Audio Processing*, vol. 37, no. 10, pp. 1495–1503, 1989.

A. Narayanan and D. Wang, *Ideal Ratio Mask Estimation Using Deep Neural Networks for Robust Speech Recognition*, Proceedings of the IEEE International Conference on Acoustics, Speech and Signal Processing (ICASSP), pp. 7092–7096, 2013.

R. Peharz and F. Pernkopf, *On Linear and Mixmax Interaction Models for Single Channel Source Separation*, Proceedings of the IEEE International Conference on Acoustics, Speech and Signal Processing (ICASSP), pp. 249–252, 2012.

M. H. Radfar, A. H. Banihashemi, R. M. Dansereau, and A. Sayadiyan, Nonlinear minimum mean square error estimator for mixture-maximisation approximation, *Electronics Letters*, vol. 42, no. 12, pp. 724–725, 2006.

B. Raj, R. Singh, and T. Virtanen, *Phoneme-Dependent NMF for Speech Enhancement in Monaural Mixtures*, Proceedings of the International Conference on Spoken Language Processing (INTERSPEECH), pp. 1217–1220, 2011.

A. M. Reddy and B. Raj, *Soft Mask Estimation for Single Channel Speaker Separation*, Proceedings of Workshop on Statistical and Perceptual Audio Processing, 2004.

R. Rose, E. Hofstetter, and D. Reynolds, Integrated models of signal and background with application to speaker identification in noise, *IEEE Transactions on Speech and Audio Processing*, vol. 2, no. 2, pp. 245–257, 1994.

S. T. Roweis, *Factorial Models and Refiltering for Speech Separation and Denoising*, Proceedings of the European Conference on Speech Communication and Technology (Eurospeech), pp. 1009–1012, 2003.

P. Smaragdis, Convolutive speech bases and their application to supervised speech separation, *IEEE/ACM Transactions on Audio, Speech, and Language Processing*, vol. 15, no. 1, pp. 1–12, 2007.

M. R. Spiegel, *Mathematical Handbook of Formulas and Tables*, McGraw Hill, 1992.

S. Srinivasan, N. Roman, and D. L. Wang, Binary and ratio time frequency masks for robust speech recognition, *Speech Communication*, vol. 48, no. 11, pp. 1486–1501, 2006.

N. Sturmel and L. Daudet, *Iterative Phase Reconstruction of Wiener Filtered Signals*, Proceedings of the IEEE International Conference on Acoustics, Speech and Signal Processing (ICASSP), pp. 101–104, 2012.

N. Sturmel and L. Daudet, Informed source separation using iterative reconstruction, *IEEE/ACM Transactions on Audio, Speech, and Language Processing*, vol. 21, no. 1, pp. 178–185, 2013.

N. Sturmel, L. Daudet, and L. Girin, *Phase-Based Informed Source Separation for Active Listening of Music*, Proceedings of the International Conference on Digital Audio Effects (DAFx-12), 2012.

A. Varga and R. Moore, *Hidden Markov Model Decomposition of Speech and Noise*, Proceedings of the IEEE International Conference on Acoustics, Speech and Signal Processing (ICASSP), pp. 845–848 vol. 2, 1990.

E. Vincent, R. Gribonval, and C. Fevotte. Performance measurement in blind audio source separation, *IEEE Transactions on Audio, Speech, and Language Processing*, vol. 14, no. 4, pp. 1462–1469, 2006.

E. Vincent, J. Barker, S. Watanabe, J. Le Roux, F. Nesta, and M. Matassoni, *The Second CHiME Speech Separation and Recognition Challenge: Datasets, Tasks and Baselines*, Proceedings of the IEEE International Conference on Acoustics, Speech and Signal Processing (ICASSP), pp. 126–130, 2013.

T. Virtanen, Monaural sound source separation by nonnegative matrix factorization with temporal continuity and sparseness criteria, *IEEE Transactions on Audio, Speech, and Language Processing*, vol. 15, no. 3, pp. 1066–1074, 2007.

T. Virtanen, J. F. Gemmeke, B. Raj, and P. Smaragdis, Compositional models for audio processing: Uncovering the structure of sound mixtures, *IEEE Signal Processing Magazine*, vol. 32, no. 2, pp. 125–144, 2015.

D. Wang, On ideal binary mask as the computational goal of auditory scene analysis, in *Speech Separation by Humans and Machines* (ed. P. Divenyi), Kluwer, pp. 181–197, 2005.

Y. Wang, *Supervised Speech Separation Using Deep Neural Networks*, PhD dissertation, Ohio State University, 2015.

D. Wang and G. J. Brown, *Computational Auditory Scene Analysis: Principles, Algorithms and Applications*, Wiley-IEEE Press, 2006.

Y. Wang and D. Wang, *Cocktail Party Processing via Structured Prediction*, Proceedings of Conference on Neural Information Processing Systems (NIPS), pp. 224–232, 2012.

Y. Wang and D. Wang, Towards scaling up classification-based speech separation, *IEEE Transactions on Audio, Speech, and Language Processing*, vol. 21, no. 7, pp. 1381–1390, 2013.

Y. Wang and D. Wang, *A Deep Neural Network For Time-Domain Signal Reconstruction*, Proceedings of the IEEE International Conference on Acoustics, Speech and Signal Processing (ICASSP), pp. 4390–4394, 2015.

Y. Wang, A. Narayanan, and D. Wang, On training targets for supervised speech separation, *IEEE/ACM Transactions on Audio, Speech, and Language Processing*, vol. 22, no. 12, pp. 1849–1858, 2014.

M. K. Watanabe and P. Mowlaee, *Iterative Sinusoidal-Based Partial Phase Reconstruction in Single-Channel Source Separation*, Proceedings of the International Conference on Spoken Language Processing (INTERSPEECH), pp. 832–836, 2013.

F. Weninger, H. Erdogan, S. Watanabe, E. Vincent, J. Le Roux, J. R. Hershey, and B. Schuller, *Speech Enhancement with LSTM Recurrent Neural Networks and its Application to Noise-Robust ASR*, Proceedings of the International Conference on Latent Variable Analysis and Signal Separation (LVA/ICA), pp. 91–99, 2015.

D. S. Williamson, Y. Wang, and D. Wang, Reconstruction techniques for improving the perceptual quality of binary masked speech, *The Journal of the Acoustical Society of America*, vol. 136, no. 2, pp. 892–902, 2014.

D. S. Williamson, Y. Wang, and D. Wang, Complex ratio masking for monaural speech separation, *IEEE/ACM Transactions on Audio, Speech, and Language Processing*, vol. 24, no. 3, pp. 483–492, 2016.

M. Wohlmayr and F. Pernkopf, Model-based multiple pitch tracking using factorial HMMs: Model adaptation and inference, *IEEE/ACM Transactions on Audio, Speech, and Language Processing*, vol. 21, no. 8, pp. 1742–1754, 2013.

M. Zöhrer, R. Peharz, and F. Pernkopf, Representation learning for single-channel source separation and bandwidth extension, *IEEE/ACM Transactions on Audio, Speech & Language Processing*, vol. 23, no. 12, pp. 2398–2409, 2015.

6

Phase-Aware Speech Quality Estimation

Pejman Mowlaee

Graz University of Technology, Graz, Austria

6.1 Chapter Organization

While various studies have been conducted on evaluating and analyzing the correlation of instrumental measures with subjective listening tests, the importance of phase in speech quality estimation has not been adequately studied. Therefore, in this chapter we address the importance of phase in speech quality estimation, in particular when the spectral phase of the signal is modified. Through several examples, we will show how the conventional instrumental measures are limited to the assessment of magnitude-only modification schemes.

As the conventional measures mostly rely on the distortion metrics applied to the spectral magnitude of a reference (desired) signal and the estimated (processed) signal, it is not clear how reliable the conventional measures are when used to predict the performance of a phase-aware speech processing application (for two examples, see Chapters 4 and 5). Therefore, we seek to address the following two fundamental questions:

- How well do the conventional instrumental measures predict the performance of signal enhancement methods where both spectral amplitude and phase are modified?
- Is it possible to improve the reliability of the speech quality prediction by considering new measures?

In this chapter, to address the first question we will present a correlation analysis between conventional instrumental measures and subjective listening tests for phase-aware speech enhancement. To answer the second question, we suggest a list of candidates for new *phase-aware* instrumental metrics, where spectral distortions due to both spectral magnitude and phase are taken into account.

6.2 Introduction: Speech Quality Estimation

The estimation of speech quality is highly important, in particular in designing a speech communication system that is expected to deliver a predefined level of quality of service.

Single Channel Phase-Aware Signal Processing in Speech Communication: Theory and Practice, First Edition.
Pejman Mowlaee, Josef Kulmer, Johannes Stahl, and Florian Mayer.

During the development of a new algorithm, subjective listening tests provide important information, in particular at intermediate steps to optimize some key parameters. It is important to note here that in this chapter, by the term "speech quality" we infer two individual concepts: perceived speech quality and speech intelligibility, which are different aspects used to estimate the quality of a speech communication design. Therefore, in the following, we will first provide a general definition of speech quality and later will continue our presentation on its two aspects: *perceived* quality and *speech intelligibility*.

6.2.1 General Definition of Speech Quality

The term *quality* is a subjective definition that differs from one user to another, depending on prior knowledge and experiences. The overall impression for the listener of how *good* the quality of speech is defines the overall speech quality. However, the term *good* is subjective and depends solely on the listener. The large difference in interpretation of the speech signal's quality makes quality estimation a multi-disciplinary task. Sound quality is characterized by a multitude of features; to name a few: naturalness, clarity, brightness, intelligibility, noisiness, artifacts. For an overview of speech quality estimation see Quackenbush *et al.* (1988), Möller *et al.* (2011), Kleijn *et al.* (2015), and Chapter 11 of Loizou (2013).

To narrow down our focus in the current chapter, we will concentrate on speech quality estimation for performance evaluation of speech signal enhancement algorithms. In fact, while the character of sound quality is a difficult topic with a variety of ways to measure it, there is unified agreement on the two important attributes in estimating the quality of a speech signal as the outcome of a communication system:

- perceived speech quality
- speech intelligibility.

The perceived quality describes how natural the speech sounds to a human, while speech intelligibility quantifies the expected number of correctly recognized words by the human auditory system.

It is quite common to measure the two aforementioned attributes by conducting subjective listening tests. In order to achieve conclusive results in this way, a large number of test subjects and test data are required. This demanding prerequisite makes subjective listening tests quite sophisticated. However, subjective listening tests are inevitable for quantifying the character of quality, or to estimate the speech quality achievable by an algorithm.

There exist several subjective listening tests aiming at measuring different aspects of perceived quality or intelligibility of speech. For example, the *comparison category rating* (CCR) test (ITU 1996) aims to quantify the preference between two sounds in terms of how much one is preferred over another, or whether there is no preference between them.

Another alternative for a subjective listening test is the *multiple stimuli with hidden reference and anchor* (MUSHRA) test (MUSHRA 2001), which is targeted at quantifying the intermediate difference between excerpts at a large scale, and therefore presents how much a sound is preferred in absolute and relative terms, or among the rest of the methods.

Barker and Cooke (2007) proposed a routine to evaluate the intelligibility of speech signals in noise. Their study was focused on the GRID corpus which consists of 34 speakers (18 male and 16 female). Each utterance in this corpus follows a command-like structure (Cooke *et al.* 2006). They reported the recognition accuracy of human listeners measured for colors, letters, and digits in the utterances from GRID corpus. The specific task was to recognize the letter and the digit spoken by the target speaker, who always said "white."

In all these subjective tests, it is highly recommended and important to keep the test conditions both unbiased and consistent during tests conducted with a large enough number of participants. Satisfying such a requirement is cumbersome if not impossible. As a consequence, performing a reasonably reliable subjective listening test is a tedious and time-consuming procedure. This is why it is highly important to find an instrumental predictor of speech quality with a high correlation to subjective listening results with human listeners. Having access to such reliable instrumental predictors of quality immensely simplifies the development and optimization of speech processing algorithms.

6.2.2 Speech Quality Estimators: Amplitude, Phase, or Both?

In Figure 6.1, we consider three different ways of evaluating the performance of a speech enhancement method:

- evaluating distortions of the spectral magnitude
- evaluating distortions of the spectral phase
- evaluating distortions of both the spectral amplitude and phase.

We refer to the first approach, amplitude only, as the conventional instrumental measures where the focus is on the spectral amplitude distortion, and hence no phase distortion is taken into account. In the second group, we consider the alternative extreme scenario where only phase information is taken into account while the amplitude is kept fixed from the clean input (for a list of the proposed phase-aware measures, see Section 6.5). In the third group, both the amplitude and phase distortions are taken into account. For example, it has been reported that phase is a useful complement to spectrally based audio quality estimation in sinusoidal modeling to properly reflect the artifacts encountered due to amplitude and phase distortion (Hollomey *et al.* 2016).

In this chapter, we will explain each aspect of speech quality estimation in more detail, and draw several important conclusions following the correlation analysis performed between the objective scores and the subjective listening results.

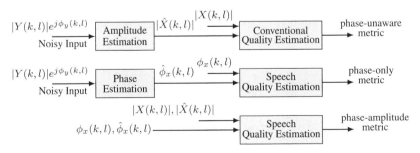

Figure 6.1 Block diagram for (top) conventional instrumental metric without using phase information, (middle) phase-only instrumental metrics, and (bottom) joint amplitude and phase metric.

6.3 Conventional Instrumental Metrics for Speech Quality Estimation

For a list of existing methods of speech quality assessment, see Table 6.1. In the following, we present a review of each category.

6.3.1 Perceived Quality

The requirements of an instrumental metric has to fulfill depend on the application that it is designed for. In the category of objective measures for predicting the perceived speech quality, three groups are generally recognized:

- speech codec based methods
- SNR-based methods
- source separation based methods.

The metrics in the first group were previously proposed for the evaluation of speech codecs for encoding/decoding in digital speech transmission. In speech coding, we aim to minimize the number of bits needed to represent the source signal (speech). This should be accomplished without degradation in its transparency at the receiver side, measured by a perceptual distortion measure. In particular, transparent speech coding quality is defined in terms of the statistics of the spectral distortion, where the number of outliers should fulfill certain constraints (So and Paliwal 2007). In speech codec based metrics, it is assumed that the speech signal follows an autoregressive process modeled by linear prediction at short time frames. Further, these metrics consider psychoacoustic

Table 6.1 List of instrumental metrics to predict perceived speech quality.

Abbreviation	Objective Measure
PESQ (Rix *et al.* 2001)	Perceptual evaluation of speech quality
GSNR (Deller *et al.* 2000)	Global SNR
SSNR (Deller *et al.* 2000)	Segmental SNR
fwSNR (Tribolet *et al.* 1978)	Frequency weighted SNR
SDR (Vincent *et al.* 2006)	Signal-to-distortion ratio
SIR (Vincent *et al.* 2006)	Signal-to-interference ratio
SAR (Vincent *et al.* 2006)	Signal-to-artifact ratio
MSE (Kay 1993)	Mean square error of phase
IS (Itakura and Saito 1970)	Itakura–Saito distance
CEPS (Kitawaki *et al.* 1988)	Cepstral distance
LLR (Gray and Markel 1976)	Log-likelihood ratio
GDD (Gaich and Mowlaee 2015a)	Group delay deviation
IFD (Gaich and Mowlaee 2015a)	Instantaneous frequency deviation
PD (Gaich and Mowlaee 2015a)	Phase deviation
UnHPSNR (Gaich and Mowlaee 2015b)	Unwrapped harmonic phase SNR
UnRMSE (Gaich and Mowlaee 2015b)	Unwrapped root mean square estimation error

models to take into account human perception. These assumptions helped to develop quality measures that fit to speech codec evaluation. A well-known example in this category is the *perceptual evaluation of speech quality* (PESQ; Rix *et al.* 2001), recommended by the ITU. Other conventional measures in this category are the cepstral distance (CEPS; Kitawaki *et al.* 1988), the Itakura–Saito (IS) distance (Itakura and Saito 1970), and the log-likelihood ratio (LLR; Gray and Markel 1976).

The distortion types quantified by speech codec based measures are of a different nature compared to those occurring in speech enhancement. Therefore, the measures are not necessarily reliable in quantifying the noise reduction capability or improved signal enhancement achieved. For example, the quantization error in a coded signal introduced by the *code-excited linear prediction* (CELP) codec is different from the musical noise or speech distortions which are of interest in the course of evaluating a speech enhancement method. Hu and Loizou (2008) reported a Pearson's correlation of $\rho = 0.78$ between PESQ and subjective listening results. This study, however, only included enhancement methods relying on amplitude modification only where the noisy spectral phase is applied directly for signal reconstruction, hence no phase-aware processing is taken into account. Therefore, it was not revealed how reliable these measures are when applied to predict the achievable performance of phase-aware speech enhancement methods.

In the group of SNR-based methods, the measures compare the reference and the estimated signals in terms of their temporal or spectral representations. The SNR-based distortions penalize any difference between the signal waveforms, sample by sample, in the sense of possible introduced delays or gain changes. Such overall signal modifications are not necessarily reliable predictors for the subjective perceived quality. Therefore, two SNR-based considerations have been suggested, taking into account the perceptual relevance with subjective listening results:

1) Short-time frames where the speech signals are stationary should be used in the calculation of the final metric. This is justified by the finding that *segmental SNR* (SSNR) has a higher correlation with human perception than the time domain version, that is, the *global SNR* (GSNR; Deller *et al.* 2000).
2) Perceptual weights should be applied to improve the correlation of SNR-based measures. As an example, the *frequency weighted SNR* (fwSNR; Tribolet *et al.* 1978) measure was shown to have a higher correlation to subjective tests than GSNR or SSNR, where no frequency weighting is taken into account.

The SNR-based measures have commonly been used in the evaluation of speech processing methods. However, as they emphasize a sample-by-sample comparison, their direct use can be misleading when the spectral phase is also changed. Considering a time domain SNR measure, one sample delay would change all the squared errors, showing even a negative value and hence indicating a poor quality, while the two signals are just a pure time-shifted version of each other but perceptually the same. As another example, the use of fwSNR only emphasizes the difference in the spectral amplitude; therefore, a signal with enforced synthetic harmonicity shows an improved quality, while it perceptually suffers from a certain buzzyness (see Experiment 6.2 in Section 6.7.2 for further details).

Finally, the last group originates from the blind source separation community, where the idea is to recover all of the underlying signals from an observed mixture. Unlike the

SNR-based measures in the second group, Vincent *et al.* proposed to factorize the degradation into three types:

1) target distortion
2) interference
3) algorithmic artifacts.

Projection matrices were derived in order to find the residual error after projecting the error signal into each of these subspaces, leading to three *blind source separation evaluation* (BSS EVAL) measures (Vincent *et al.* 2006):

1) signal-to-distortion ratio (SDR)
2) signal-to-interference ratio (SIR)
3) signal-to-artifact ratio (SAR).

The measures have been standardized for the source separation community; their reliability in predicting quality (Mowlaee *et al.* 2012a; Erdogan *et al.* 2015) and word recognition accuracy (Weninger *et al.* 2015) have been reported. Emiya *et al.* (2011) extended the BSS EVAL measures to *perceptual evaluation for audio source separation* (PEASS) measures. The PEASS measures have a higher correlation with subjective listening tests than BSS EVAL. Again, all these measures neglect the phase information, and are hence required to account for some proper time alignment due to the possible delays in the estimated signal. This requirement is important for the sake of a fair performance evaluation among separation algorithms where the spectral phase is modified.

6.3.2 Speech Intelligibility

Unlike perceived speech quality, speech intelligibility quantifies the number of correctly recognized speech units, for example phonemes or words. For a list of speech intelligibility measures see Table 6.2. In general, the conventional intelligibility metrics rely on defining a distortion function that approximates the intelligibility in one of the following ways:

- as a weighted sum of signal-to-noise ratios calculated at different frequency bands[1] (ANSI 1997; Kates and Arehart 2005; Sauert and Vary 2010),
- Incorporating the mutual information between the modified signal and the reference signal (Taghia and Martin 2014; Jensen and Taal 2014);
- relying on auditory models emphasizing the importance of each frequency band;
- applying octave-level normalization and compression (Taal *et al.* 2011a).

Table 6.2 shows a list of the conventional speech intelligibility metrics. The main principle followed by all these conventional measures is to approximate the SNR in frequency bands to estimate the speech intelligibility. Assuming that the SNR is representative for speech audibility, those frequency bands where the SNR is positive are concluded to have the highest contribution to speech intelligibility. Therefore, a weighted sum of SNRs

---

1 Depending on the selected weight and SNR calculation, several different intelligibility measures have been developed (as discussed in this section).

Table 6.2 List of speech intelligibility metrics.

Abbreviation	Objective Measure
SII (ANSI 1997)	Speech intelligibility index
CSII (Kates and Arehart 2005)	Coherence SII
NCM (Holube and Kollmeier 1996)	Normalized covariance measure
SNRloss (Ma and Loizou 2011)	Loss in signal-to-noise ratio
STOI (Taal *et al.* 2011a)	Short-time objective intelligibility measure
MIKNN (Taghia and Martin 2014)	Mutual information k-nearest neighbor
SIMI (Jensen and Taal 2014)	Speech intelligibility prediction based on mutual information
DAU (Christiansen *et al.* 2010)	Dau metric for intelligibility prediction
GDD (Gaich and Mowlaee 2015a)	Group delay deviation
IFD (Gaich and Mowlaee 2015a)	Instantaneous frequency deviation
PD (Gaich and Mowlaee 2015a)	Phase deviation
UnHPSNR (Gaich and Mowlaee 2015b)	Unwrapped harmonic phase SNR
UnRMSE (Gaich and Mowlaee 2015b)	Unwrapped root mean square estimation error

defined across all frequency bands is often used to approximate the speech intelligibility given by (Loizou 2013, Ch. 13):

$$\mathrm{SI} = \sum_{k=1}^{K} W_k \mathrm{SNR}_k, \tag{6.1}$$

where K is the total number of frequency bands with $k \in [1, K]$, W_k approximates the band importance function normalized to fulfill $\sum_k W_k = 1$, and SI denotes the predicted speech intelligibility score. The band importance function W_k should be designed to incorporate the characteristics of human auditory perception, applying psychoacoustic principles.

Intelligibility can be considered as the correct recognition of words (Allen 1994). Allen later found a relationship between the *articulation index* (AI) of Fletcher's model (Fletcher and Galt 1950) and the resulting automatic speech recognition errors. The articulation index quantifies the contribution of each frequency with regard to the intelligibility of the speech signal and is calculated using the intensities of the speech and noise in the frequency domain. As one way to approximate the articulation index, Kryter (1962) introduced the *speech intelligibility index* (SII) measure as the weighted sum of the AI, weighted according to a band importance function. The approximation formula in (6.1) has been the origin of many speech intelligibility predictors, including: speech intelligibility index (SII; ANSI 1997), SNRloss (Ma and Loizou 2011), *coherence SII* (CSII; Kates and Arehart 2005), and the *normalized covariance measure* (NCM; Holube and Kollmeier 1996). The measures differ in their method of calculating the SNR and the weighting function selected therein. For example, the SII measures take two psychoacoustic aspects into account when calculating the effective band SNR (also called audibility function): (i) equivalent masking spectrum as level distortion factor, and (ii) hearing threshold level. The resulting SII is then given by:

$$SII = \sum_{k=1}^{K} AI_k(SNR_{k_i}),$$ (6.2)

where AI_k accounts for the band importance function defined for the frequency band $k \in [1, K]$ selected from a table (ANSI 1997), and K is the total number of frequency bands, for example, $K = 18$. Speech and noise spectral levels in decibels are then used to approximate SNR_k as the SNR at the kth frequency subband. In 6.2, $AI_k(SNR_k)$ refers to the weighting, often called the band audibility function, and the SII measure lies in the range $0 \leq SII \leq 1$. An SII value above 0.75 is interpreted as high intelligibility, while a value below 0.45 indicates poor speech intelligibility. The SII has been widely used in speech communication, for example, for optimizing the near-end listening enhancement (Sauert and Vary 2010), as a predictor of speech intelligibility (Fletcher and Galt 1950), and pre-enhancement (Taal and Jensen 2013) to derive a convex optimization principle.

As an alternative to SII, CSII is similar to the SII measure but differs by replacing the SNR term with SDR_k, called the signal-to-distortion ratio. It is calculated at the kth critical band filtered by a bank of rounded-exponential filters providing approximations to the auditory filters.[2] Instead of the SNR_k used in SII, a *signal-to-distortion ratio* (SDR) based on the *magnitude-squared coherence function* (MSC) is used. SDR is calculated at short-term spectra as follows:

$$SDR_k = \frac{|\gamma_k|^2}{1 - |\gamma_k|^2},$$ (6.3)

where $|\gamma_k|$ is the MSC function given by

$$|\gamma_k|^2 = \frac{|\sum_l X(k,l)Y^*(k,l)|^2}{|\sum_l X(k,l)|^2 |\sum_l Y(k,l)|^2}.$$ (6.4)

The CSII metric is represented in three levels, depending on the RMS level of the magnitude of a segment relative to the overall RMS level:

- $CSII_h$: high level, above the overall RMS level;
- $CSII_m$: mid level, between 0 dB and 10 dB below the overall RMS level;
- $CSII_l$: Low level, between 10 dB and 30 dB below the overall RMS level.

Figure 6.2 illustrates the RMS level assigned to the three levels of CSII for a speech signal.

Holube and Kollmeier (1996) presented the NCM intelligibility measure that relies on the calculation of the covariance between the input and output signals in terms of their envelope in the frequency bands. The resulting SNR calculated in this domain is then limited to the range between -15 dB and 15 dB. A linear mapping is used in order to compute the resulting *transmission index* (TI), based on the SNR values at frequency band k by $TI_k = \frac{SNR_k + 15}{30}$. Finally, the overall NCM index is obtained by averaging over all frequency bands:

$$NCM = \sum_{k=1}^{K} W_k TI_k.$$ (6.5)

---

2 For details, see Table 1 in ANSI (1997).

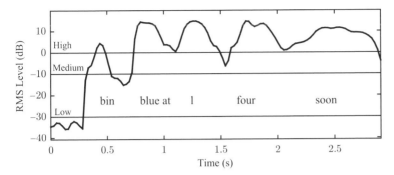

Figure 6.2 Segmentations of a TIMIT sentence based on its RMS levels to the low, mid, and high regions used in the CSII method.

Later, Ma and Loizou (2011) proposed an intelligibility metric relying on the loss of SNR at different critical bands, which is calculated between the reference (clean) and a noise-suppressed (modified) speech signal. The resulting SNRloss measure predicts the speech intelligibility loss caused by the enhancement algorithm; for the lth frame and the kth frequency band it is given by

$$\mathrm{SNRloss}(k, l) = \mathrm{SNR}_X(k, l) - \mathrm{SNR}_{\hat{X}}(k, l), \tag{6.6}$$

where X and $\hat{X}$ are the STFT magnitudes of the input reference signal and the enhanced signal.

Christiansen *et al.* (2010) proposed an intelligibility predictor relying on the auditory model first presented in Dau *et al.* (1997). Such a psychoacoustic model is a successful tool in developing their instrumental measure, relying on the calculation of the similarity between the modified signal and the reference at some intermediate representation levels. Averaged cross-correlation coefficients were calculated at frames of length 20 ms and were classified in three levels: low, mid, and high. Using a weighted sum of these three level scores, the final DAU metric for intelligibility prediction is obtained.

Last, but not least, several attempts have been made to approximate speech intelligibility by using the mutual information concept. The mutual information expresses the similarity between two messages, for example, the one transmitted through a communication medium and the one interpreted at the receiver end (listener). The mutual information can be used as a natural measure to assess the intelligibility (Kleijn *et al.* 2015). Kleijn *et al.* presented a unified framework for optimizing speech intelligibility in noisy environments using the mutual information concept. It is important to note that the mutual information criterion can be applied at different levels of the signal: sequence of uttered words, sequence of states of the auditory system, or on the whole message. For example, Taghia and Martin (2014) proposed the *mutual information k-nearest neighbour* (MIKNN) objective intelligibility measure, relying on estimating the mutual information between the processed speech and a reference clean speech signal in terms of their corresponding temporal envelope. As another example, Jensen and Taal (2014) proposed using the mutual information to estimate the speech intelligibility as a function of a mean square error criterion, assuming that the intelligibility is monotonically related to the mutual information defined between the amplitude envelope of the clean and the processed signals in their critical bands. The measure was termed as *speech intelligibility prediction based on mutual information* (SIMI). Through their results, the authors confirmed the reduced speech intelligibility by the amplitude-only modification speech enhancement method reported earlier by Loizou and Kim (2011).

Most of the speech intelligibility measures rely, in their calculations, on the global statistics captured at a sentence level. In contrast, Taal *et al.* (2011a) proposed the *short-time objective intelligibility* (STOI) measure relying on shorter time segments (386 ms). The STOI measure was shown to achieve a high correlation with the listening test results and has been widely used for the performance evaluation of speech enhancement methods in the literature.

6.4   Why Phase-Aware Metrics?

We address the question in two examples. In our first example we demonstrate the importance of phase information in the resulting speech intelligibility. Controlled modification of the phase by adding noise to the phase at harmonics demonstrates the importance of phase information for highly intelligible reconstructed signals. As our second experiment, we highlight the deficiency of the existing instrumental metrics for a reliable prediction of the speech intelligibility of an enhanced signal whose spectral phase is also modified.

6.4.1   Phase and Speech Intelligibility

We conduct an experiment for speech intelligibility evaluation; the results are shown in Figure 6.10. The speech intelligibility shares much with the spectral phase information and its temporal fine structure (see, e.g., Alsteris and Paliwal 2006). We use STFTPI (Krawczyk and Gerkmann 2014) as the phase modification approach, replacing the noisy phase at signal reconstruction, while the noisy spectral amplitude is used directly. For comparison purposes, we also include the oracle scenario, where the clean phase spectrum only is available and used for signal reconstruction together with the noisy spectral amplitude. This outcome highlights the importance of phase-only information in signal reconstruction, and should be interpreted as the upper-bound performance for the phase-only scenario. For more details see Experiment 6.2 in Section 6.7.

6.4.2   Phase and Perceived Quality

In this section we want to emphasize the limits and the caution required to be taken into account when interpreting the predicted perceived quality by conventional instrumental measures (as these metrics solely rely on the spectral amplitude). The phase-insensitive characteristics of the existing metrics lead to a mismatch with the phase-aware signal processing modifications that are applied in phase-aware enhancement algorithms and algorithms used for evaluating them.

Here, we consider the interaction of the predicted perceived quality and the applied phase estimation method. Details on this experiment are presented in Experiment 6.1. This experiment shows a counter example for quality prediction of phase-aware speech enhancement where we consider a phase-only enhancement scenario. The phase-only enhancement scenario refers to when the noisy spectral phase is replaced by an estimated one while the noisy spectral amplitude is not modified. In this example, we consider the estimated phase provided by STFTPI Krawczyk and Gerkmann (2014). Figure 6.9 shows the outcomes for the unprocessed (noisy) speech, clean reference, and phase-only reconstructed speech signals. The representations are shown in terms of spectrogram (top) and phase variance (bottom) plots. The spectrogram comparison emphasizes the impact of phase as a result of signal reconstruction, and hence is an indirect consequence of the selected phase spectrum. The phase variance, in contrast, has been reported as a reliable predictor of voice quality for synthesized speech (Koutsogiannaki *et al.* 2014). Further, Kates (2000) reported that the phase variance is a reliable metric to estimate the noise and distortion in an automatic gain control in hearing aids, where the conventional magnitude-squared coherence function leads to errors due to the sensitivity to the system gain fluctuations.

6.5 New Phase-Aware Metrics

In this section, we consider phase-only measures to study the possibility of evaluating modified speech signals through quantifying the amount of phase distortion. The reliability of the proposed measures will be justified by subjective listening tests. The measures considered here are divided into two groups: (i) STFT-based and (ii) harmonic-based methods. The STFT-based group consists of the following methods: *group delay deviation* (GDD), *instantaneous frequency deviation* (IFD), *unwrapped mean square error* (UnMSE), and *phase deviation* (PD). The harmonic-based group consists of *unwrapped harmonic phase SNR* (UnHPSNR) and unwrapped root mean square estimation error (UnRMSE).

6.5.1 Group Delay Deviation

Details of the group delay representation and its successful usage for different speech processing applications were given in Chapter 2. Group delay de-emphasizes the cyclic wrapping issue, and represents a useful harmonic structure in speech processing. The group delay deviation was proposed in Stark and Paliwal (2009) and has been used for phase estimation[3] from speech mixtures (Mowlaee *et al.* 2012b) or noisy speech (Mowlaee and Saeidi 2014). These findings motivated Gaich and Mowlaee (2015b) to propose a distortion metric relying on group delay, given as follows:

$$d_{\text{GDD}} = \frac{2}{LK} \sum_{l=1}^{L} \sum_{k=1}^{K} (\cos(\tau_x(k,l)) - \cos(\hat{\tau}_x(k,l)))^2, \tag{6.7}$$

where $\tau_x(k,l)$ and $\hat{\tau}_x(k,l)$ denote the group delay of the reference signal and the modified signal for the kth frequency and the lth frame, respectively, L is the total number of frames, and K is the number of frequency bins. To make the distance metric invariant to modulo 2π and to avoid wrong error calculations due to the periodicity of phase, in the DD metric the cosine function is applied. A similar treatment was employed for phase-based estimators studied in Lagrange and Marchand (2007), as well as in deriving an estimator for the spectral phase in Cohen (2005).

---

3 For more details on geometry-based phase estimators, see Chapter 3.

6.5.2 Instantaneous Frequency Deviation

For more details on the instantaneous frequency, we refer to Chapter 2. Alsteris and Paliwal (2007) first introduced the concept of *instantaneous frequency deviation* (IFD). The IFD representation is useful for short-time spectral phase analysis and is defined as:

$$\mathrm{IFD}_{\phi_x}(k, l) = (\phi_x(k, l) - \phi_x(k, l - 1)) - \frac{2\pi S k}{K}, \tag{6.8}$$

with frame shift S. The IFD carries information about the vocal tract excitation (Portnoff 1981). It is also a useful representation for pitch estimation or automatic speech recognition as it resolves the formant frequencies (Alsteris and Paliwal 2007). Gaich and Mowlaee (2015a) proposed a metric based on IFD given by:

$$d_{\mathrm{IFD}} = \frac{1}{LK} \sum_{l=1}^{L} \sum_{k=1}^{K} (\cos(\mathrm{IFD}_{\phi_x}(k, l)) - \cos(\mathrm{IFD}_{\hat{\phi}_x}(k, l)))^2. \tag{6.9}$$

The criterion was also used to resolve the ambiguity in phase estimation for single-channel speech enhancement (Mowlaee and Saeidi 2014).

6.5.3 Unwrapped MSE

While the *mean square error* (MSE) criterion has been widely used in the literature of estimation theory in order to evaluate the correctness of the amplitude, frequency, and phase estimated values for a signal in noise (see, e.g., Kay 1993), the cyclic property of phase renders the squared criterion incomplete and in fact misleads in interpreting performance and analyzing errors. In order to circumvent the phase wrapping problem in evaluation of phase estimation errors, a cosine function can be applied, resulting in an MSE measure in the unwrapped domain, given by (Gaich and Mowlaee 2015a; Gaich and Mowlaee 2015a,b):

$$d_{\mathrm{UnMSE}} = \frac{1}{LK} \sum_{l=1}^{L} \sum_{k=1}^{K} (\cos(\phi_x(k, l) - \hat{\phi}_x(k, l)))^2. \tag{6.10}$$

6.5.4 Phase Deviation

In the second experiment in Chapter 1, we studied the impact of phase modification in speech enhancement proposed by Vary (1985). Vary first defined phase deviation as the difference between the noisy phase and the clean phase given by

$$\phi_{\mathrm{dev}}(k, l) = \phi_y(k, l) - \phi_x(k, l). \tag{6.11}$$

The geometric representation of the phase deviation concept is shown in Figure 6.3. The concept was later employed in Gustafsson *et al.* (2002) for joint noise reduction and echo cancellation, and in Mowlaee and Saeidi (2013)for phase estimation. In Gaich and Mowlaee (2015a), the PD concept was used to define a new distortion metric:

$$d_{\mathrm{PD}} = \frac{1}{LK} \sum_{l=1}^{L} \sum_{k=1}^{K} (\cos(\phi_{\mathrm{dev}}(k, l)) - \cos(\hat{\phi}_{\mathrm{dev}}(k, l)))^2. \tag{6.12}$$

We note that all the derived phase-aware measures discussed here (PD, IFD, GD, and UnMSE) are bounded in the interval $[0, 4]$.

Figure 6.3 Geometric representation for the single-channel speech enhancement problem, showing noisy, clean, and noise spectra denoted by $Y(k, l)$, $X(k, l)$, and $D(k, l)$, respectively. The phase deviation $\phi_{\mathrm{dev}}(k, l)$ is defined as the phase difference between the clean phase $\phi_x(k, l)$ and the noisy phase $\phi_y(k, l)$.

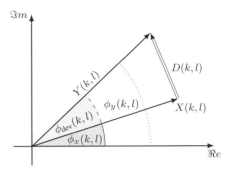

6.5.5 UnHPSNR and UnRMSE

The phase-aware measures discussed so far all rely on the STFT phase spectrum. Gaich and Mowlaee (2015b) proposed two new measures focused on quantifying the estimation error of harmonic phase. The measures are the *unwrapped root mean square estimation error* (UnRMSE),

$$\mathrm{UnRMSE} = 10\log_{10}\left(\frac{1}{L} \sum_{l=1}^{L} \sqrt{\frac{\sum_{h=1}^{H} |X(h, l)|^2 (\Psi_x(h, l) - \hat{\Psi}_x(h, l))^2}{\sum_{h=1}^{H} |X(h, l)|^2}} \right), \tag{6.13}$$

and *unwrapped harmonic phase SNR* (UnHPSNR),

$$\mathrm{UnHPSNR} = 10\log_{10}\left(\frac{1}{L} \sum_{l=1}^{L} \frac{\sum_{h=1}^{H} |X(h, l)|^2}{\sum_{h=1}^{H} |X(h, l)|^2 (1 - \cos(\Psi_x(h, l) - \hat{\Psi}_x(h, l)))} \right), \tag{6.14}$$

both measured in decibels, where $X(h, l)$ is the spectral amplitude sampled at harmonic h and frame index l, and $\Psi_x(h, l)$ and $\hat{\Psi}_x(h, l)$ are the corresponding unwrapped harmonic phase spectra. The unwrapped phase spectra are provided after subtracting the linear phase part, following pitch-synchronous segmentation using the phase decomposition principle proposed in Degottex and Erro (2014). In the definition of UnRMSE, the choice of phase variance without the mean was motivated by recent findings on the importance of phase variance for voice quality assessment (Koutsogiannaki *et al.* 2014) and the negligible impact of the phase mean (Degottex and Erro 2014). The weighting by the spectral amplitude $X(h, l)$ emphasizes the phase variance error at harmonics, which is arguably perceptually most important.

6.6 Subjective Tests

To support the improvements predicted by the instrumental metrics, in the following we present several subjective listening experiments obtained when phase-aware speech enhancement is applied. Three tests were conducted:

- *comparison category rating* (CCR) test
- *multiple stimuli with hidden reference and anchor* (MUSHRA) test
- speech intelligibility test.

The subjective listening tests were performed in a quiet room using AKG K-240 studio headphones. Fourteen listeners participated in the tests, all selected from the audio listening experts panel at the Graz University of Technology.

6.6.1 CCR Test

For the CCR test, Mowlaee and Kulmer followed the instructions in ITU (1996) recommended to assess speech processing that either degrades or improves the quality of the speech. A seven-rating scale was used ranging from -3 to 3: "(-3) much worse," "(-2) worse," "(-1) slightly worse," "(0) no difference," "(1) slightly better," "(2) better," and "(3) much better." The utterances presented to each participant were randomly selected from two male and female speakers, 20 from each mixed with white and babble noise from NOISEX-92 (Varga *et al.* 1992) at 0, 5, and 10 dB SNRs. The noisy files (unprocessed, UP) were processed by four speech-enhancement algorithms: *conventional* (C: MMSE-LSA; Ephraim and Malah 1984), *conventional + clean phase* (C + clean), *conventional + phase-enhanced* (C + PE; Kulmer *et al.* 2014), and *phase-aware* (PA; Mowlaee and Saeidi 2013). Including the unprocessed files, a total of 1500 speech files were collected for the correlation analysis. The results of the instrumental metrics were averaged over all utterances for each method and SNR.

The averaged *comparison mean opinion score* (CMOS) showed a preference of slightly better to better, indicating that the combination of enhanced amplitude using MMSE-LSA (Ephraim and Malah 1985) and enhanced spectral phase using the temporal smoothing method outperforms the noisy phase scenario.

In another study, Maly and Mowlaee (2016) conducted a CCR test to study the importance of linear phase and unwrapped smoothed phase in signal reconstruction for speech enhancement. Their results showed that an improved harmonic phase is sufficient rather than a clean STFT phase, as they perform very closely. The comparison between subjective and objective results revealed that a proper unwrapped smoothed phase and linear phase sufficed to obtain a phase-enhanced signal.

6.6.2 MUSHRA Test

In order to evaluate the intermediate difference in quality between phase-modified signals, Gaich and Mowlaee (2015a) performed a subjective listening test following the MUSHRA standard recommended in MUSHRA (2001). Fifty utterances from the GRID corpus (Cooke *et al.* 2006) composed of male and female speakers downsampled to 8 kHz were used. Following the MUSHRA standard (MUSHRA 2001), they also included a hidden reference as well as an anchor (defined as the 2.5 kHz lowpass filtered reference signal).

Figure 6.4 shows the *mean opinion scores* (MOS) and 95% confidence intervals differentiated in terms of noise type and SNR. For all noise types and SNRs, similar rankings were observed where the PA method performed best followed by the C + clean phase method. The PE method was ranked as the third with a short gap with respect to the C method. The UP, as expected, had the lowest ranking. T-tests were conducted to justify the significance of these rankings. Except between the C and the PE methods, all other rankings were significant with respect to each other with a p-value of $p < 0.05$. However, C + PE outperforms C significantly for white noise at SNR = 0 dB as well as at SNR = 5 dB with $p = 0.077$ (see Table 6.3 for comparison between LSA + PE versus LSA).

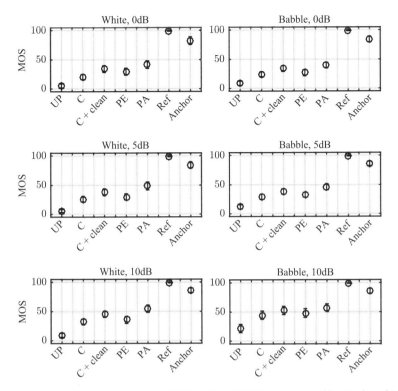

Figure 6.4 Mean opinion scores (MOS) of the MUSHRA test inspired by Gaich and Mowlaee (2015a). White noise (left) and babble noise (right) shown for 11 participants. The results are grouped into (top) low SNR = 0 dB, (middle) mid SNR = 5 dB, and (bottom) high SNR = 10 dB.

Table 6.3 Subjective evaluation results for intelligibility test, reported in percentages comparing LSA and LSA + PE methods.

LSA+PE	LSA
0.7714 ± 0.1026	0.6429 ± 0.1349

Another observation is that the overall perceptual quality improvement between the unprocessed and the PE method is more pronounced in white noise than babble noise. In contrast, the quality improvement achieved by method C is approximately the same for both noise types. This is due to the fact that blind f_0 estimation (Gonzalez and Brookes 2014) works better in white noise, leading to a more accurate estimated phase for the PE method (Kulmer *et al.* 2014; Kulmer and Mowlaee 2015a).

6.6.3 Statistical Analysis

For performance evaluation, we used the same approach as presented in Taal *et al.* (2009), where the Pearson's correlation coefficient (ρ), the normalized root-mean-square error (σ), and Kendall's tau (τ_k) were used as three figures of merit. We followed the same procedure proposed in Taal *et al.* (2009) by applying two different logistic functions to account for the nonlinear relationship between the objective and subjective scores.

To further evaluate the reliability of the results, another correlation analysis at each SNR as well as on the whole data labeled as (SNRs, Noise) was conducted, with the results shown in Table 6.4. In Table 6.4, only metrics with a correlation $\rho > 0.8$ are shown. The metrics which fulfill this constraint are the proposed phase-aware metrics as well as PESQ among the conventional metrics. To obtain the last row (labeled as Mean), the correlation coefficients are averaged. The PD metric shows the highest correlation with perceived quality ($\rho = 0.92$) on average, followed by PESQ ($\rho = 0.9$). Furthermore, the PD metric performs most reliably as it shows a more stable ρ across noise types. Following the correlation analysis in Steiger (1980), the ranking of the proposed PD metric on (SNRs, Noise), is significant compared to global SNR, SSNR, fwSNR, BSS EVAL, MSE, LLR, IFD, and CEPS.

6.6.4 Speech Intelligibility Test

Gaich and Mowlaee (2015b) studied speech intelligibility evaluation for phase-aware speech enhancement. Fifty sentences from GRID were used at 8 kHz sampling frequency with SNRs selected at 0 dB and 5 dB. Four enhancement methods were considered: *Conventional* (C: MMSE-LSA; Ephraim and Malah 1984), *conventional + STFT phase improvement* (C + STFTPI; Krawczyk and Gerkmann 2014), *conventional + phase enhanced* (C + PE; Kulmer and Mowlaee 2015a), and *phase-aware* (PA; Mowlaee and Saeidi 2013). Together with the unprocessed speech signals (UP), 1000 speech files were used in the analysis. Scores of the instrumental metrics were obtained by averaging over all utterances for each method and SNR.

For the speech intelligibility tests, Gaich and Mowlaee followed the principle and standard recommended in Barker and Cooke (2007) to conduct a subjective listening test for speech intelligibility performance evaluation. The participants were instructed to choose the right color, letter, and number at each presented utterance selected from the GRID corpus. The test was organized in four blocks according to noise type and decreasing SNR. Within each block, for each method four randomly selected utterances were presented to the participants, where the order of the methods itself was also randomized. To check the reliability of the participants, clean reference utterances were included. Participants were excluded from the final statistical analysis once their intelligibility score for the clean utterances showed a value lower than the noisy utterances (as the clean signal is supposed to have the highest performance). This led to ten participants in the test. A training session was provided.

Figure 6.5 shows the intelligibility scores and 95% confidence intervals differentiated in terms of noise type and SNR. An overall improvement in intelligibility is more pronounced in babble rather than in white noise except for C + STFTPI and C + PE in the babble 0 dB case, where the noisy phase is replaced with an estimated one at the signal reconstruction stage. These methods work in the harmonic domain and therefore need a reliable estimation of the fundamental frequency, which is an erroneous task in adverse noise. This fact also supports the improved intelligibility attained by C + PE compared to C in white than babble noise, verified with the results reported in Gaich and Mowlaee (2015a), where for C + PE against C better quality improvement was shown in the white noise case. The iterative amplitude–phase estimation method is labeled as PA, and outperforms all the other methods in every scenario. For the babble 0 dB case this improvement is significant. A two-proportion z-test was conducted to calculate the

Table 6.4 Statistical analysis of the top performing perceived quality metrics for different noise types and SNRs, averaged over both SNRs and noise types.

	PESQ (Rix et al. 2001)			GDD (Gaich and Mowlaee 2015a)			IFD (Gaich and Mowlaee 2015a)			PD (Gaich and Mowlaee 2015a)		
	ρ	σ	τ_k	ρ	σ	τ_k	ρ	σ	τ_k	ρ	σ	τ_k
Noise = babble	0.87	0.07	0.77	0.84	0.07	−0.70	0.84	0.07	−0.68	**0.91**	0.06	−0.77
Noise = white	0.86	0.07	0.79	**0.96**	0.04	−0.87	0.89	0.07	−0.81	0.94	0.05	−0.89
SNR = 0 dB	0.91	0.05	0.91	0.81	0.07	−0.69	0.87	0.06	−0.87	**0.92**	0.05	−0.87
SNR = 5 dB	**0.95**	0.04	0.91	0.83	0.07	−0.73	0.92	0.05	−0.69	0.93	0.05	−0.87
SNR = 10 dB	**0.93**	0.05	0.91	0.91	0.06	−0.78	0.88	0.07	−0.82	0.90	0.06	−0.82
(SNRs, Noise)	0.86	0.07	0.73	0.87	0.07	−0.71	0.86	0.07	−0.71	**0.91**	0.06	−0.77
Mean	0.90	0.06	0.84	0.87	0.06	−0.75	0.88	0.07	−0.76	**0.92**	0.06	−0.83

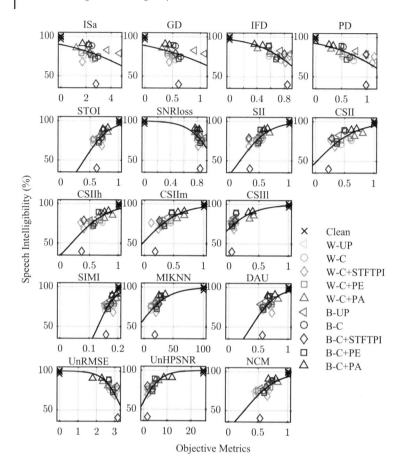

Figure 6.5 Correlation results for speech intelligibility measures with the subjective listening results.

significance at the 95% confidence level. PA also showed significant improvement in comparison to C except for the case of babble noise at SNR 5 dB case. Finally, the method C degrades the intelligibility of the noisy input, confirming the previous observations made in Loizou and Kim (2011) and Taal *et al.* (2014).

6.6.5 Evaluation of Speech Quality Measures

The Pearson's correlation coefficient (ρ), the normalized root-mean-square error (σ), and Kendall's tau (τ_k) were used to evaluate the prediction power of the presented instrumental metrics. To account for the nonlinear relationship between objective and subjective scores, a logistic function was applied.

Figure 6.5 shows the mapped scatter plot for each intelligibility measure for a combined SNR scenario at 0 dB and 5 dB for both white and babble noise scenarios. The following observations can be made: The *linear predictive coding* (LPC) based measures CEPS, LLR, and IS together with the proposed phase-aware measure GD show a low correlation ($\rho < 0.8$) due to the underestimation of the unprocessed conditions. The other phase-aware measures, IFD, PD, and MSE, show a moderate correlation

$(0.8 < \rho < 0.9)$ but are not capable of reaching the top performing intelligibility measures. The harmonic-based phase-aware measures, UnRMSE and UnHPSNR, show the highest correlation for white noise and a reasonably high correlation for babble noise. Although UnHPSNR is calculated in the same domain asUnRMSE, it shows less correlation. We explain this observation by two facts: (i) in the unwrapped domain, the exact difference between the unwrapped phases $(\Psi - \hat{\Psi})$ is more reliable than a metric which uses an additional cosine term; and (ii) UnHPSNR computes scores weighted by spectral amplitude at harmonics and neglects frames capturing spectral transitions, which are argued to be important for speech intelligibility (Greenberg 2006). Although UnRMSE and UnHPSNR perform reliably in the separated noise scenarios, they are not capable of predicting the subjective results in the combined noise scenario due to a consistent under-prediction of the babble noise outcomes and a clear over-prediction of the STFTPI method at 0 dB for babble noise. This significant overestimation was observed for every instrumental measure and explains the degraded performance for the combined noise scenario in comparison to the performance reported for each individual noise scenario.

CSII-based measures predict reliably in both noise scenarios. For babble noise and the overall scenario, CSIIm is the top performing metric. This is supported by an observation by Kates and Arehart (2005) that the mid-level CSII contains much information on envelope transients. The CSIIh score mainly represents vowel contributions and does not distinguish between, for example, /v/ and /b/, which differ in their onsets but share the same vowel phoneme resulting in a lower correlation than CSIIm. These transients are mostly affected by phase enhancement, which, as a consequence, led CSIIm to be a reliable intelligibility predictor for phase-enhanced speech.

The mutual information based measures MIKNN and SIMI, reported to be reliable predictors for speech intelligibility in speech enhancement, in this experiment showed a moderate prediction performance. We emphasize that these measures compute a score on a long-term statistic which may cause inaccurate results using the short sentences of GRID ($\sim$1.5 seconds). Finally, the DAU measure correlates moderately, as reported in Taal *et al.* (2014).

The correlation results combining both noise types were in general at a lower level. This is due to two facts: first, no metric was capable of predicting the intelligibility loss by STFTPI at 0 dB for babble noise as seen in Figure 6.5; and second, an over-estimation of intelligibility occurred in white noise, in particular for the PA speech enhancement outcome predicted by UnRMSE, leading to an intermediate correlation score.

Table 6.5 summarizes the top performing instrumental measures that exhibit $\rho \geq 0.8$ for intelligibility prediction in each noise scenario. IFD, PD, MSE together with STOI, the conventional SII, CSII, CSIIh, and CSIIm constitute the subset. On average, the highest correlation is obtained by CSIIm ($\rho = 0.91$), which outperforms all other measures, followed by CSII and IFD. High performance of CSII-based measures for single-channel amplitude-only enhanced speech was already reported in Ma *et al.* (2009), Taal and Jensen (2013), and Taal *et al.* (2011a). The IFD and MSE measures show better performance than STOI in all scenarios. Although the PD measure exhibited the lowest correlation coefficient in this subset of the top performing intelligibility measures, it shows a comparable performance to CSIIm in terms of τ and furthermore reveals the most stable results across noise types, a property that was already observed in the perceived quality evaluation reported in the previous section (see Table 6.5).

Table 6.5 Statistical analysis of the top performing speech intelligibility metrics for different noise and SNRs, averaged over both SNRs and noise.

	White			Babble			Both			Mean		
	ρ	σ	τ_k	ρ	σ	τ_k	ρ	σ	τ_k	ρ	σ	τ_k
IFD	0.90	**0.04**	−0.74	0.89	0.07	−0.75	0.81	0.07	−0.64	0.87	0.06	−0.71
PD	0.83	0.05	**−0.80**	0.83	0.08	−0.81	0.80	0.08	−0.67	0.82	0.07	−0.76
MSE	0.90	**0.04**	0.71	0.89	0.07	0.81	0.80	0.07	0.68	0.86	0.06	0.73
STOI	0.89	**0.04**	0.68	0.84	0.08	0.81	0.80	0.07	−0.63	0.84	0.07	0.70
SII	0.91	**0.04**	0.77	0.81	0.09	0.75	0.80	0.07	0.66	0.84	0.07	0.73
CSII	0.91	**0.04**	0.77	0.90	0.06	0.78	0.84	0.07	**0.74**	0.88	0.06	0.76
CSIIh	0.84	0.05	0.77	0.89	0.07	0.81	0.83	0.07	0.72	0.85	0.06	**0.77**
CSIIm	**0.92**	**0.04**	**0.80**	**0.94**	**0.05**	**0.84**	**0.86**	**0.06**	0.68	**0.91**	**0.05**	0.77

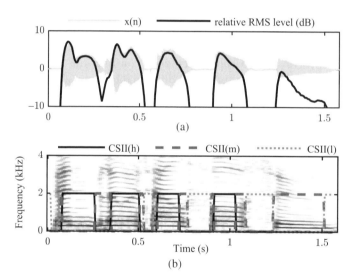

Figure 6.6 Speech signal (top) and spectrogram (bottom) of the utterance "bin blue at I four soon."

In order to investigate the reason why CSII performs as a reliable predictor of speech intelligibility, we demonstrate a time domain signal which is labeled in order to highlight which signal regions contribute to which CSII level. The results are shown in Figure 6.6. It can be seen that the CSIIm is concentrated on the transients, known to have the highest contribution in speech intelligibility.

6.7 Experiments

We consider the following experiments:

1) Experiment 6.1: Impact of phase modifications on speech quality
2) Experiment 6.2: Phase and perceived quality estimation

3) Experiment 6.3: Phase and speech intelligibility estimation
4) Experiment 6.4: Evaluating the phase estimation accuracy.

These experiments will highlight the limits of the conventional metrics and their unreliability in predicting the perceived quality or intelligibility achieved by a phase-aware speech processing algorithm. They will also demonstrate the importance of phase information in speech quality estimation in terms of perceived quality and speech intelligibility.

6.7.1 Experiment 6.1: Impact of Phase Modifications on Speech Quality

In this section, we demonstrate how phase information impacts on the perceived quality or intelligibility of a speech signal. To this end, we consider a clean speech signal, and decompose its STFT phase $\phi_x(k, l)$ into the underlying components, using harmonic phase decomposition (Degottex and Erro 2014), as reviewed in Chapter 2. The unwrapped phase outcome at harmonic h and frame l denoted by $\Psi(h, l)$ is given by subtracting the linear phase component,

$$\Psi(h, l) = \psi(h, l) - \psi_{\text{lin}}(h, l), \tag{6.15}$$

defining $\psi_{\text{lin}}(h, l)$ as the linear phase component. To study how noise affects the spectral phase, we add a random noise phase variable $\psi_{\text{n}}(h, l) \sim \mathcal{U}[-\pi, \pi[$, where $\mathcal{U}[a, b[$ denotes a uniform distribution within the range $[a, b[$. The randomness of the added noise is controlled by a scaling parameter $\alpha_0 \in [0, 1]$; the modified noise-corrupted unwrapped phase is given by

$$\hat{\Psi}_{\text{mod}}(h, l) = \Psi(h, l) + \alpha_0 \psi_{\text{n}}(h, l). \tag{6.16}$$

This modified unwrapped phase is used to obtain an instantaneous phase by adding the linear phase part. The distorted unwrapped phase is transformed back into the STFT domain by modifying the frequency bins within the mainlobe width of the analysis window[4] denoted by N_p,

$$\hat{\phi}_{x_{\text{mod}}}(\lfloor h\omega_0 K \rfloor + i, l) = \hat{\Psi}_{\text{mod}}(h, l) + \psi_{\text{lin}}(h, l) \quad \forall i \in [-N_p/2, N_p/2], \tag{6.17}$$

with K as the DFT size and $\omega_0 = 2\pi f_0/f_s$. The complex spectrum is then constructed using the clean spectral amplitude $X(k, l)$, together with the modified phase given as:

$$X_{\text{mod}}(k, l) = X(k, l)e^{j\hat{\phi}_{x_{\text{mod}}}(k, l)}. \tag{6.18}$$

This modified complex spectrum is eventually used to reconstruct speech using the overlap-add procedure. Figure 6.7 shows the impact of added noise to the unwrapped harmonic phase for different values of $\alpha_0 = \{0, 0.3, 0.6, 1\}$. The influence of such phase modifications is visualized in terms of phase variance (top) and spectrogram (bottom) representations. The following observations can be made:

- A large modification in the unwrapped phase changes the phase variance, and masks the harmonic structure observed in the clean speech spectrogram (first column). This impacts the speech intelligibility of the reconstructed signal captured by the instrumental measures CSII, DAU, and STOI, as shown in Figure 6.8.

---

4 See Chapter 3 for more details.

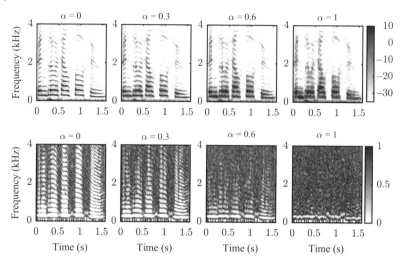

Figure 6.7 Circular variance and spectrogram for the clean phase signal, phase-modified signal $(\alpha_0 = 0, 0.3, 0.6, 1)$ and randomized phase $(\alpha_0 = 1)$.

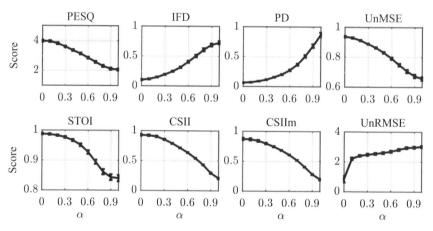

Figure 6.8 Mean objective scores for the best performing instrumental measures evaluated over 50 GRID utterances corrupted by phase distortions controlled by $\alpha_0 \in [0, 1]$. The results are shown for (top) quality measures (PESQ, IFD, PD, UnMSE) and (bottom) intelligibility measures (STOI, CSII, CSIIm, UnRMSE).

- Although the spectral amplitude is not modified, the change in the unwrapped phase results in harsh sound quality, visualized by the widened bandwidth of the harmonics. The effect of phase modification on the frequency distortion can be explained by the fact that the time or frequency derivatives of the phase, known as instantaneous phase and group delay, are characterized by their smooth changes. Through randomizing the unwrapped phase, the continuity of these representations is lost.

Finally, we study the speech quality outcome of speech synthesized using the modified unwrapped phase. The intelligibility performance versus α_0 is shown in the lower panel in Figure 6.8. The results are reported in terms of intelligibility predictors (STOI, CSII, UnRMSE) and quality predictors (PESQ, IFD, PD, and UnMSE), averaged over

50 utterances from five speakers (two male, three female) and for different values of $\alpha_0 \in [0, 1]$. From Figure 6.8 we can see that:

- Speech intelligibility does not completely break down by phase-only modification at harmonics. In particular, for $\alpha_0 \leq 0.4$ only 5% degradation in STOI is experienced, while for $\alpha_0 > 0.5$ the intelligibility drops about 50% in CSII.
- Due to the imperfect harmonic decomposition, no single measure achieves its best score even for $\alpha_0 = 0$. The STOI measure shows insensitivity to the introduced phase distortions, which results in an over-prediction of speech intelligibility for high values of α_0.
- Both CSII and PD measures showed the most sensitivity, spanning a wide dynamic range with respect to the scores, with no saturation towards $\alpha_0 \to 1$. The over-sensitivity of the CSII-based measures with respect to the phase was also reported in Taal *et al.* (2011b).

6.7.2 Experiment 6.2: Phase and Perceived Quality Estimation

We consider a noisy speech signal composed of a female utterance corrupted in white noise at 0 dB SNR. The noisy signal is produced by contaminating a female utterance saying "she had your dark suit in greasy wash water all year," selected from TIMIT (Garofolo *et al.* 1993). The clean speech is corrupted with white noise at 10 dB signal-to-noise ratio. The utterance consists of stop consonants and fricatives, reported to be important for speech intelligibility (Liu *et al.* 1997). The outcome after applying the STFTPI phase estimation method (Krawczyk and Gerkmann 2014) is shown in the middle panel, versus the clean reference signal on the left and the noisy observation shown on the right. Comparing the spectrograms, group delay, and phase variance patterns, we see that the STFTPI method (Krawczyk and Gerkmann 2014) entirely removes the signal components between harmonics. The implementation for this experiment is found in *Exp6_2.m*, available in the *PhaseLab Toolbox* (see the appendix).

The results are shown in Figure 6.9. The outcomes for PESQ and fwSNR as two conventional predictors of perceived quality are shown in the top panels. The following observations can be made:

- Both PESQ and fwSNR measures predict a certain improved perceived quality for STFTPI. These results confirm the earlier observations made in Krawczyk and Gerkmann (2014). However, the phase-enhanced outcome suffers from some buzzyness, due to the artifacts visible when comparing the harmonic structure in the phase-enhanced signal versus the clean original signal (a similar trend is visible in the group delay plot). These artifacts were reported in Krawczyk and Gerkmann (2014), Patil and Gowdy (2014), and Gerkmann (2014).
- The contradiction between the predicted improved quality by conventional metrics (PESQ and fwSNR) and the buzzy signal quality by STFTPI motivated us to address the fundamental questions that we started with in this chapter. The distortion is not captured by the conventional measures such as PESQ and fwSNR; therefore, these measures and their corresponding results and ranking need to be used with caution and are under question.

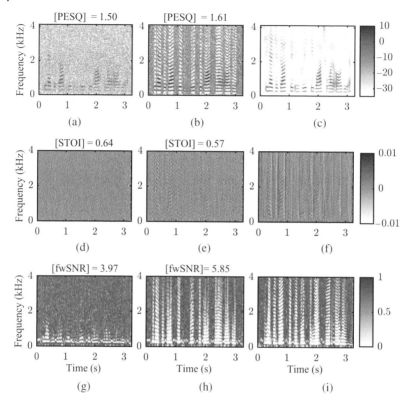

Figure 6.9 Experiment 6.2: Noisy (left), STFTPI (Krawczyk and Gerkmann 2014; middle), and clean (right) speech signals. Results shown as spectrogram (top), group delay (middle), and phase variance (bottom). The predicted quality using PESQ and frequency-weighted SNR are shown for each outcome at the top of each panel.

6.7.3 Experiment 6.3: Phase and Speech Intelligibility Estimation

We repeat the previous experiment to study the intelligibility performance obtained from a phase enhancement method. For this, we include the clean phase scenario that highlights the information conveyed for speech intelligibility when the noisy phase is replaced with the clean phase while the noisy spectral amplitude is used unaltered. Further, we include the STFTPI method (Krawczyk and Gerkmann 2014) for comparison purposes. The implementation for this experiment is found in *Exp6_3.m*, available in the *PhaseLab Toolbox* (see the appendix).

The results are shown in Figure 6.10, illustrating the outcomes for clean speech (clean amplitude and clean phase), noisy, STFTPI, and clean phase signals. The intelligibility performance instrumentally predicted by STOI (Taal *et al.* 2011a) and *coherence speech intelligibility index* (CSII; Kates and Arehart 2005) are shown above the panels. The following observations can be made:

- The phase-only information brings a significantly improved speech intelligibility predicted by both instrumental measures used. In particular, the harmonic structure in the clean signal is partially recovered when using the phase-only information.

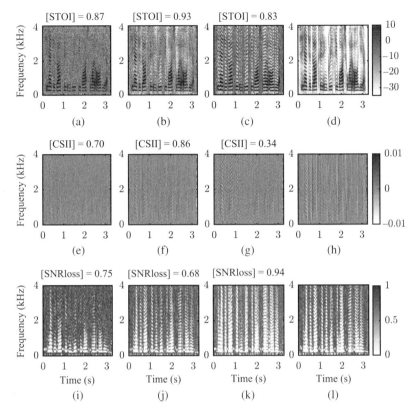

Figure 6.10 Results shown for clean, noisy, estimated and clean phase: (top) spectrogram, (middle) group delay, (bottom) phase variance. Speech intelligibility outcome predicted by STOI and CSII, for a phase-enhanced signal using STFTPI.

This observation emphasizes the importance of phase estimation in noise for speech enhancement, and further confirms the earlier observations in Alsteris and Paliwal (2006).

- The STFTPI degrades the speech intelligibility by 10% compared to the unprocessed speech measured by the STOI and CSII metrics. This example confirms that both STOI and CSII predict the degraded intelligibility performance due to harmonized phase modification.

6.7.4 Experiment 6.4: Evaluating the Phase Estimation Accuracy

As our last experiment, we compare some of the phase-aware measures proposed in this chapter. We evaluate the performance using two groups: (i) conventional measures PESQ, STOI, and CSII; and (ii) the phase-aware ones, PD, unRMSE, and unHPSNR. The results are shown for a male utterance taken from a GRID utterance saying "bin blue at e six please" corrupted in white noise at 5 dB. We consider the following phase modification scenarios:

A noisy (unprocessed)
B STFTPI (Krawczyk and Gerkmann 2014)

C maximum *a posteriori* (MAP; Kulmer and Mowlaee 2015b)
D clean STFT phase (upper bound).

The implementation for this experiment is found in *Exp6_4.m*, available in the *PhaseLab Toolbox* (see the appendix). The results are reported in Table 6.6.

We emphasize the two conditions required for an evaluation criterion to be reliable:

1) It reaches its lowest predicted value for the unprocessed (noisy) signal and its largest one for the clean spectral phase, known as the upper bound for phase estimation performance.
2) It predicts an improved perceived quality and does not over- or under-estimate it due to some irrelevant modification with regard to the clean signal (e.g., artifacts and buzzyness).

From the results reported in Table 6.6, the following observations can be made:

• Both PESQ and PD metrics predict high perceived quality scores for the clean phase and the lowest scores for noisy. However, PESQ overestimates the quality for STFTPI, where spurious harmonics are introduced leading to a buzzy quality. In contrast, PD penalizes these artifacts (harmonization) by predicting a worse quality (note that the lower the PD the better the estimated quality).
• For speech intelligibility, the CSII predicts the speech intelligibility quite reliably, concluding that both clean and estimated phase have a positive contribution to an improved intelligibility of speech in noise. In contrast, a buzzy quality due to artifacts and harmonization is predicted as a degraded speech intelligibility score.
• The phase-aware measures, unHPSNR and unRMSE, predict the best results for clean phase and the worst one for the STFTPI, where the phase structure and estimates are distorted. It is important to note that the noisy phase at strong signal components (mostly at low frequencies) is still similar to the clean phase, therefore a reasonable value is predicted, still lower than the clean phase as the upper bound.

Table 6.6 Results for different phase modification scenarios in terms of conventional and phase-aware measures. The phase modification methods are: (A) noisy (unprocessed), (B) STFTPI, (C) maximum *a posteriori* (MAP), (D) clean STFT phase (upper bound).

Metrics	A	B	C	D
PESQ	1.55	1.69	1.66	2.51
STOI	0.79	0.74	0.83	0.91
PD	0.77	1.15	0.60	0.21
CSII	0.45	0.21	0.48	0.73
UnRMSE	2.32	3.60	2.18	0.26
UnHPSNR	4.13	0.40	4.98	8.44

6.8 Summary

In this chapter we investigated the important issue of speech quality estimation where both spectral amplitude and phase are modified. Unlike the conventional instrumental quality predictors in the literature that rely on amplitude-only modification, our studies showed that phase-aware measures are capable of providing similar or even better prediction and correlation with subjective listening results. The limits and misleading interpretation of results obtained by measures originally recommended for amplitude-only scenarios were displayed and carefully investigated through several proof-of-concept examples. The correlation analysis for the instrumental measures consisting of conventional and new phase-aware candidates were demonstrated with conclusions drawn.

A future direction to investigate is to address a combination of the amplitude and phase spectra in a new instrumental measure and to compare its correlation compared to phase-only or the current amplitude-only intelligibility or quality measures that have been used for a long time in single-channel speech enhancement.

References

J. B. Allen, How do humans process and recognize speech?, *IEEE Transactions on Speech and Audio Processing*, vol. 2, no. 4, pp. 567–577, 1994.

L. D. Alsteris and K. K. Paliwal, Further intelligibility results from human listening tests using the short-time phase spectrum, *Speech Communication*, vol. 48, no. 6, pp. 727–736, 2006.

L. D. Alsteris and K. K. Paliwal, Short-time phase spectrum in speech processing: A review and some experimental results, *Digital Signal Processing*, vol. 17, no. 3, pp. 578–616, 2007.

ANSI S3.5-1997, *Methods for Calculation of the Speech Intelligibility Index*, American National Standards Institute, 1997.

J. Barker and M. Cooke,Modelling speaker intelligibility in noise, *Speech Communication*, vol. 49, no. 5, pp. 402–417, 2007.

C. Christiansen, M. S. Pedersen, and T. Dau, Prediction of speech intelligibility based on an auditory preprocessing model, *Speech Communication*, vol. 52, no. 7–8, pp. 678–692, 2010.

I. Cohen, Relaxed statistical model for speech enhancement and a priori SNR estimation, *IEEE Transactions on Speech and Audio Processing*, vol. 13, no. 5, pp. 870–881, 2005.

M. Cooke, J. Barker, S. Cunnigham, and X. Shao, An audio-visual corpus for speech perception and automatic speech recognition, *The Journal of the Acoustical Society of America*, vol. 120, pp. 2421–2424, 2006.

T. Dau, B. Kollmeier, and A. Kohlrausch, Modeling auditory processing of amplitude modulation. I. Detection and masking with narrow-band carriers, *The Journal of the Acoustical Society of America*, vol. 102, no. 5, pp. 2892–2905, 1997.

G. Degottex and D. Erro, A uniform phase representation for the harmonic model in speech synthesis applications, *EURASIP Journal on Audio, Speech, and Music Processing—Special Issue: Models of Speech—In Search of Better Representations*, 2014.

J. R. J. Deller, J. H. L. Hansen, and J. G. Proakis, *Discrete-Time Processing of Speech Signals*, Wiley-IEEE Press, 2000.

V. Emiya, E. Vincent, N. Harlander, and V. Hohmann, Subjective and objective quality assessment of audio source separation, *IEEE Transactions on Audio, Speech, and Language Processing*, vol. 19, no. 7, pp. 2046–2057, 2011.

Y. Ephraim and D. Malah, Speech enhancement using a minimum mean square error short-time spectral amplitude estimator, *IEEE Transactions on Acoustics, Speech, and Signal Processing*, vol. 32, no. 6, pp. 1109–1121, 1984.

Y. Ephraim and D. Malah, Speech enhancement using a minimum mean-square error log-spectral amplitude estimator, *IEEE Transactions on Acoustics, Speech, and Signal Processing*, vol. 33, no. 2, pp. 443–445, 1985.

H. Erdogan, J. R. Hershey, S. Watanabe, and J. Le Roux, *Phase-Sensitive and Recognition-boosted Speech Separation using Deep Recurrent Neural Networks*, Proceedings of the IEEE International Conference on Acoustics, Speech and Signal Processing (ICASSP), 2015.

H. Fletcher and R. H. Galt, The perception of speech and its relation to telephony, *The Journal of the Acoustical Society of America*, vol. 22, no. 2, pp. 89–151, 1950.

A. Gaich and P. Mowlaee, *On Speech Quality Estimation of Phase-Aware Single-Channel Speech Enhancement*, Proceedings of the IEEE International Conference on Acoustics, Speech and Signal Processing (ICASSP), pp. 216–220, 2015a.

A. Gaich and P. Mowlaee, *On Speech Intelligibility Estimation of Phase-Aware Single-Channel Speech Enhancement*, Proceedings of the International Conference on Spoken Language Processing (INTERSPEECH), pp. 2553–2557, 2015b.

J. S. Garofolo, L. F. Lamel, W. M. Fisher, J. G. Fiscus, D. S. Pallett, and N. L. Dahlgren, *DARPA TIMIT Acoustic Phonetic Continuous Speech Corpus CDROM*, National Institute of Standards and Technology (NIST), 1993.

T. Gerkmann, *MMSE-Optimal Enhancement of Complex Speech Coefficents With Uncertain Prior Knowledge of The Clean Speech Phase*, Proceedings of the IEEE International Conference on Acoustics, Speech and Signal Processing (ICASSP), pp. 4511–4515, 2014.

S. Gonzales and M. Brookes, PEFAC: A pitch estimation algorithm robust to high levels of noise, *IEEE/ACM Transactions on Audio, Speech, and Language Processing*, vol. 22, no. 2, pp. 518–530, 2014.

A. Gray and J. Markel, Distance measures for speech processing, *IEEE Transactions on Acoustics, Speech, and Signal Processing*, vol. 24, no. 5, pp. 380–391, 1976.

S. Greenberg, A multi-tier theoretical framework for understanding spoken language, in *Listening to Speech: An Auditory Perspective* (eds S. Greenberg and W. A. Ainsworth), Lawrence Erlbaum Associates, pp. 411–433, 2006.

S. Gustafsson, R. Martin, P. Jax, and P. Vary, A psychoacoustic approach to combined acoustic echo cancellation and noise reduction, *IEEE Transactions on Speech and Audio Processing*, vol. 10, no. 5, pp. 245–256, 2002.

C. Hollomey, D. Moore, D. Knox, W. O. Brimijoin, and W. Whitmer, Using phase information to improve the reconstruction accuracy in sinusoidal modeling, *Audio Engineering Society*, 2016.

I. Holube and B. Kollmeier, Speech intelligibility prediction in hearing-impaired listeners based on a psychoacoustically motivated perception model, *The Journal of the Acoustical Society of America*, vol. 100, no. 3, pp. 1703–1716, 1996.

Y. Hu and P. C. Loizou, Evaluation of objective quality measures for speech enhancement, *IEEE Transactions on Audio, Speech, and Language Processing*, vol. 16, no. 1, pp. 229–238, 2008.

F. Itakura and S. Saito, A statistical method for estimation of speech spectral density and formant frequencies, *Electronic Communications*, vol. 53, no. 1, pp. 36–43, 1970.

ITU, *Methods for Subjective Determination of Transmission Quality—Series P: Telephone Transmission Quality; Methods for Objective and Subjective Assessment of Quality, Recommendation P. 800*, International Telecommunication Union (ITU), 1996.

ITU, *Method for the Subjective Assessment of Intermediate Quality Level of Coding Systems*, ITU-R BS.1534-1, 2001.

J. Jensen and C. H. Taal, Speech intelligibility prediction based on mutual information, *IEEE/ACM Transactions on Audio, Speech, and Language Processing*, vol. 22, no. 2, pp. 430–440, 2014.

J. M. Kates, Cross-correlation procedures for measuring noise and distortion in AGC hearing aids, *The Journal of the Acoustical Society of America*, vol. 107, pp. 3407–3414, 2000.

J. Kates and K. Arehart, Coherence and the speech intelligibility index, *The Journal of the Acoustical Society of America*, vol. 115, no. 5, pp. 2224–2237, 2005.

S. M. Kay, *Fundamentals of Statistical Signal Processing: Estimation Theory*, Prentice-Hall PTR, 1993.

N. Kitawaki, H. Nagabuchi, and K. Itoh, Objective quality evaluation for low-bit-rate speech coding systems, *IEEE Journal on Selected Areas in Communications*, vol. 6, no. 2, pp. 242–248, 1988.

W. B. Kleijn, J. B. Crespo, R. C. Hendriks, P. Petkov, B. Sauert, and P. Vary, Optimizing speech intelligibility in a noisy environment: A unified view, *IEEE Signal Processing Magazine*, vol. 32, no. 2, pp. 43–54, 2015.

M. Koutsogiannaki, O. Simantiraki, G. Degottex, and Y. Stylianou, The Importance of Phase on Voice Quality Assessment, Proceedings of the International Conference on Spoken Language Processing (INTERSPEECH), Sept. 2014.

M. Krawczyk and T. Gerkmann, STFT phase reconstruction in voiced speech for an improved single-channel speech enhancement, *IEEE/ACM Transactions on Audio, Speech, and Language Processing*, vol. 22, no. 12, pp. 1931–1940, 2014.

K. D. Kryter, Methods for the calculation and use of the articulation index, *The Journal of the Acoustical Society of America*, vol. 34, no. 11, pp. 1689–1697, 1962.

J. Kulmer and P. Mowlaee, Phase estimation in single channel speech enhancement using phase decomposition, *IEEE Signal Processing Letters*, vol. 22, no. 5, pp. 598–602, 2015a.

J. Kulmer and P. Mowlaee, *Harmonic Phase Estimation in Single-Channel Speech Enhancement Using Von Mises Distribution and Prior SNR*, Proceedings of the IEEE International Conference on Acoustics, Speech and Signal Processing (ICASSP), pp. 5063–5067, 2015b.

J. Kulmer, P. Mowlaee, and M. A. Watanabe, *A Probabilistic Approach For Phase Estimation in Single-Channel Speech Enhancement Using Von Mises Phase Priors*, Proc. IEEE Workshop on Machine Learning for Signal Processing (MLSP), 2014.

M. Lagrange and S. Marchand, Estimating the instantaneous frequency of sinusoidal components using phase-based methods, *Audio Engineering Society*, vol. 55, no. 5, pp. 385–399, 2007.

L. Liu, H. Jialong, and P. Günther, Effects of phase on the perception of intervocalic stop consonants, *Speech Communication*, vol. 22, no. 4, pp. 403–417, 1997.

P. Loizou, *Speech Enhancement: Theory and Practice*, CRC Press, 2013.

P. C. Loizou and G. Kim, Reasons why current speech-enhancement algorithms do not improve speech intelligibility and suggested solutions, *IEEE Transactions on Audio, Speech, and Language Processing*, vol. 19, no. 1, pp. 47–56, 2011.

J. Ma and P. C. Loizou, SNR loss: A new objective measure for predicting the intelligibility of noise-suppressed speech, *Speech Communication*, vol. 53, no. 3, pp. 340–354, 2011.

J. Ma, Y. Hu, and P. Loizou, Objective measures for predicting speech intelligibility in noisy conditions based on a new band-importance function, *The Journal of the Acoustical Society of America*, vol. 125, no. 5, pp. 3387–3405, 2009.

A. Maly and P. Mowlaee, *On the Importance of Harmonic Phase Modification for Improved Speech Signal Reconstruction*, Proceedings of the IEEE International Conference on Acoustics, Speech and Signal Processing (ICASSP), pp. 584–588, 2016.

P. Mowlaee and R. Saeidi, Iterative closed-loop phase-aware single-channel speech enhancement, *IEEE Signal Processing Letters*, vol. 20, no. 12, pp. 1235–1239, 2013.

P. Mowlaee and R. Saeidi, *Time–Frequency Constraint for Phase Estimation in Single-Channel Speech Enhancement*, Proceedings of the International Workshop on Acoustic Signal Enhancement (IWAENC), pp. 338–342, 2014.

P. Mowlaee, R. Saeidi, M. G. Christensen, and R. Martin, *Subjective and Objective Quality Assessment of Single-Channel Speech Separation Algorithms*, Proceedings of the IEEE International Conference on Acoustics, Speech and Signal Processing (ICASSP), pp. 69–72, 2012a.

P. Mowlaee, R. Saeidi, and R. Martin, *Phase Estimation for Signal Reconstruction in Single-Channel Speech Separation*, Proceedings of the International Conference on Spoken Language Processing (INTERSPEECH), pp. 1548–1551, 2012b.

S. Möller, W.- Y. Chan, N. Cote, T. H. Falk, A. Raake, and M. Waltermann, Speech quality estimation: Models and trends, *IEEE Signal Processing Magazine*, vol. 28, no. 6, pp. 18–28, 2011.

S. P. Patil and J. N. Gowdy, *Exploiting the Baseband Phase Structure of the Voiced Speech for Speech Enhancement*, Proceedings of the IEEE International Conference on Acoustics, Speech and Signal Processing (ICASSP), pp. 6133–6137, 2014.

H. Pobloth, *Perceptual and Squared Error Aspects in Speech and Audio Coding*, PhD Thesis, Royal Institute of Technology (KTH), 2004.

M. R. Portnoff, Short-time Fourier analysis of sampled speech, *IEEE Transactions on Acoustics, Speech, and Signal Processing*, vol. 29, no. 3, pp. 364–373, 1981.

S. R. Quackenbush, T. P. Barnwell, and M. A. Clements, *Objective Measures of Speech Quality*, Prentice-Hall, 1988.

A. W. Rix, J. G. Beerends, M. P. Hollier, and A. P. Hekstra, *Perceptual Evaluation of Speech Quality (PESQ): A New Method for Speech Quality Assessment of Telephone Networks and Codecs*, Proceedings of the IEEE International Conference on Acoustics, Speech and Signal Processing (ICASSP), vol. 2, pp. 749–752, 2001.

B. Sauert and P. Vary, *Recursive Closed-Form Optimization of Spectral Audio Power Allocation for Near End Listening Enhancement*, ITG-Fachtagung Sprachkommunikation, 2010.

S. So and K. K. Paliwal, A comparative study of LPC parameter representations and quantisation schemes for wideband speech coding, *Digital Signal Processing*, vol. 17, no. 1, pp. 114–137, 2007.

A. P. Stark and K. K. Paliwal, *Group-Delay-Deviation Based Spectral Analysis of Speech*, Proceedings of the International Conference on Spoken Language Processing (INTERSPEECH), pp. 1083–1086, 2009.

J. H. Steiger, Tests for comparing elements of a correlation matrix, *Psychological Bulletin*, vol. 87, no. 2, pp. 245–251, 1980.

C. H. Taal and J. Jensen, *SII-Based Speech Preprocessing for Intelligibility Improvement in Noise*, Proceedings of the International Conference on Spoken Language Processing (INTERSPEECH), pp. 3582–3586, 2013.

C. H. Taal, R. C. Hendriks, R. Heusdens, and J. Jensen, An algorithm for intelligibility prediction of time–frequency weighted noisy speech, *IEEE Transactions on Audio, Speech, and Language Processing*, vol. 19, no. 7, pp. 2125–2136, 2011a.

C. H. Taal, R. C. Hendriks, R. Heusdens, and J. Jensen, An evaluation of objective measures for intelligibility prediction of time–frequency weighted noisy speech, *The Journal of the Acoustical Society of America*, vol. 130, no. 5, pp. 3013–3027, 2011b.

C. H. Taal, R. C. Hendriks, and R. Heusdens, Speech energy redistribution for intelligibility improvement in noise based on a perceptual distortion measure, *Computer Speech & Language*, vol. 28, no. 4, pp. 858–872, 2014.

C. H. Taal, R. C. Hendriks, R. Heusdens, J. Jensen, and U. Kjems, *An Evaluation of Objective Quality Measures for Speech Intelligibility Prediction*, Proceedings of the International Conference on Spoken Language Processing (INTERSPEECH), pp. 1947–1950, 2009.

J. Taghia and R. Martin, Objective intelligibility measures based on mutual information for speech subjected to speech enhancement processing, *IEEE/ACM Transactions on Audio, Speech, and Language Processing*, vol. 22, no. 1, pp. 6–16, 2014.

J. M. Tribolet, P. Noll, B. McDermott, and R. E. Crochiere, *A Study of Complexity and Quality of Speech Waveform Coders*, Proceedings of the IEEE International Conference on Acoustics, Speech and Signal Processing (ICASSP), vol. 3, pp. 586–590, 1978.

A. Varga, H. J. M. Steeneken, M. Tomlinson, and D. Jones, *The NOISEX–92 Study on the Effect of AdditiveNoise on Automatic Speech Recognition*, Technical Report, DRA Speech Research Unit, 1992.

P. Vary, Noise suppression by spectral magnitude estimation mechanism and theoretical limits, *Signal Processing*, vol. 8, no. 4, pp. 387–400, 1985.

E. Vincent, R. Gribonval, and C. Févotte, Performance measurement in blind audio source separation, *IEEE Transactions on Audio, Speech, and Language Processing*, vol. 14, no. 4, pp. 1462–1469, 2006.

F. Weninger, H. Erdogan, S. Watanabe, E. Vincent, J. Le Roux, J. R. Hershey, and B. Schuller, *Speech Enhancement with LSTM Recurrent Neural Networks and its Application to Noise-Robust ASR*, Proceedings of the International Conference on Latent Variable Analysis and Signal Separation (LVA/ICA), 2015.

7

Conclusion and Future Outlook

Pejman Mowlaee

Graz University of Technology, Graz, Austria

7.1   Chapter Organization

In this chapter, we present several future directions with potential significant break-throughs in the young field of phase-aware signal processing. A brief review of the renaissance of phase-aware processing with its decline in the past and the renewed interest in recent years is provided. We point out the series of relevant events involving phase-aware signal processing which reflect the renewed interest of researchers in the field of speech communication and eventually motivated the writing of this book. A range of applications where phase processing may successfully be incorporated in the future is presented. In particular, we describe potential applications within two categories: (i) connected research disciplines and the possible extension of the techniques and concepts learned in this book for single-channel phase-aware processing to multi-channel or other speech processing applications not fully covered here; and (ii) neighboring signal processing domains that involve signals other than speech (e.g., image, video, and biomedical signals) which could benefit from the phase-aware speech processing knowledge learned from this book.

7.2   Renaissance of Phase-Aware Signal Processing: Decline and Rise

It took researchers almost a century to start reinvestigating the misconception about the belief that the *human ear is phase-deaf*[1] and to pose questions about the importance of phase information. Only in recent years increasing interest has been devoted toward conducting research involving phase processing, whereby the phase

---

1 Observed in early studies in 1843 by Georg Simon Ohm (Ohm, 1843) and Hermann von Helmholtz (von Helmholtz, 1912).

Single Channel Phase-Aware Signal Processing in Speech Communication: Theory and Practice, First Edition.
 Pejman Mowlaee, Josef Kulmer, Johannes Stahl, and Florian Mayer.
© 2017 John Wiley & Sons, Ltd. Published 2017 by John Wiley & Sons, Ltd.

spectrum is incorporated in the signal processing solution and applied in various speech applications. Several individual steps have been taken, distributed among the different communities in speech signal processing, showing a momentum towards phase-aware signal processing. Mowlaee *et al.* (2014) organized a special conference session entitled "Phase Importance in Speech Processing Applications," focused on phase importance in speech signal processing and aiming to unify the progress recently made toward phase-based signal processing in different speech applications including speech enhancement, speech analysis/synthesis, and automatic speech and speaker recognition. The outcome of this special session was considerable shared interest shown by the many researchers who contributed to this event.

As a follow-up, in 2015 Mowlaee organized a tutorial session entitled "Phase Estimation from Theory to Practice," and presented several phase estimation methods to attract further researchers toward continuing their research in this direction. Finally, in 2016 a special issue entitled "Phase-Aware Signal Processing in Speech Communication" was organized by Mowlaee *et al.* (2015) in the Elsevier *Speech Communication* journal, where scientists submitted their recent findings in phase-aware signal processing for speech communication applications. The main goal was to unify the distributed attempts made by researchers in the diverse sub-communities in speech communication applications.

Motivated by the aforementioned series of events and publications focused on phase-aware signal processing for speech communication, and having received encouraging feedback, we decided to write the current book. In this book, the aim is to review the role of phase processing in different speech processing applications in the context of speech transmission, as shown in the blocks in Figure 1.2. We explain why phase processing has been neglected by researchers, and why they devoted more focus on estimating the spectral magnitudes in the STFT domain. We further highlight the recent increasing attention given to phase processing in different speech processing applications and communities. Early and recent techniques proposed by researchers for phase processing in speech are reviewed. Our goal is to demonstrate the usefulness of phase-based signal processing as an emerging field of research with the potential to push speech transmission toward higher quality of service and designing more robust speech communication devices, in particular in highly adverse noisy scenarios.

7.3   Directions for Future Research

It is the belief of the authors that phase-based signal processing is still in its development stage and there are several unknown aspects and unsolved problems that remain attractive for researchers. In our opinion, two main directions to follow are:

- related research disciplines, where one may involve even more disciplines related to speech communication applications into the field of phase-aware speech signal processing;
- other research disciplines, which explore the possibilities of extending the current methods available for speech processing to other research disciplines, where phase processing might play an important role.

7.3.1 Related Research Disciplines

Many questions remain open in phase-aware signal processing for speech communication that could be addressed in future studies. Below, we provide several directions to follow up the research presented in the chapters of this book.

7.3.1.1 Phase-Aware Processing for Speech and Speaker Recognition

In Chapter 1, we pointed out that the common practice in automatic speech recognition is to discard the spectral phase information and feed ASR with features derived from some energy-based features, for example, *Mel-frequency cepstral coefficient* (MFCC). Within the last two decades, however, there have been a plethora of proposals to apply some kind of phase-aware signal processing in ASR. These studies can mainly be divided into two classes: (i) front-end signal processing, and (ii) feature extraction to provide features for ASR. As some examples for front-end phase-aware processing, we refer to complex spectral subtraction (Kleinschmidt *et al.* 2011) and phase-sensitive filters (PSF; Erdogan *et al.* 2015) learned by a deep neural network (DNN). With the rapid advances in the field of neural networks, more sophisticated networks are expected that will be capable of learning the complicated structure of complex spectral coefficients and their correlation in time and frequency. A potential future direction, therefore, is to propose new deep architectures that could learn the useful structures of the underlying signal, for example, in its raw format (Tüske *et al.* 2014; Golik *et al.* 2015) or using a combination of more than one subnet, as proposed in Williamson *et al.* (2016).

Also, there have been many proposals for deriving phase-aware features to feed ASR. These proposals can be categorized into two groups: (i) those solely relying on phase-only features, for example, *chirp group delay* (CGD; Bozkurt 2005), *modified group delay* (Hegde *et al.* 2007), and *group delay power product* (Donglai and Paliwal 2004); and (ii) those demonstrating the possible combination of amplitude-derived and phase-derived features for overall improved speech recognition or speaker recognition accuracy. For example, the amplitude and phase data streams of information have been successfully combined at score or feature levels (Murty and Yegnanarayana 2006), Nakagawa *et al.* 2012). By taking the Hilbert transform, joint amplitude and phase information has been taken into account in a unified way for both speaker recognition (Sadjadi and Hansen 2011) and speech recognition (Loweimi *et al.* 2013). Taking into account more phase-aware features has the potential of improving the recognition performance in such applications, hence deserves more attention in the future.

7.3.1.2 Speech Synthesis and Speech Coding

A reliable estimate of the harmonic phase is a key factor for reasonable performance in speech coding or speech synthesis applications. It is known as a difficult problem; early proposals tried to circumvent it by using zero-phase or minimum phase spectra extracted from the magnitude spectrum (Oppenheim 1969), applying mixed phase spectra extracted from complex cepstrum (Quatieri 1979), or all-pass filtering (Hedelin 1988). Later, Pobloth and Kleijn reported the perceptual importance of phase in speech coding (Pobloth and Kleijn 1999; 2003) and studied the relevance of the squared error measure between the uncoded and coded signals. Recently, phase interpolation techniques have been successfully applied to achieve a high quality synthesized speech.

Some examples include phase-locked pitch-synchronous processing (Agiomyrgiannakis 2015), quadratic polynomial phase, and cubic phase model (McAulay and Quatieri 1986). The key idea is to obtain a smoothed phase which avoids any discontinuity that may degrade the perceived quality of the reconstructed signal. A promising future direction is then to apply a parametric harmonic phase model combined with a robust interpolation scheme so that a good balance between low bit rate constraints posed by the selected speech codec and high speech quality can be made.

Speech synthesis methods often rely on the concatenation of speech frames, either using a direct concatenation followed by a phase continuity criterion or by an overlap-add routine. In all these concatenation techniques, the continuity of the harmonic phase is of high importance. In particular, discontinuities result in noisy, reverberant speech perceived as distorted quality. Synchronous concatenation based on *glottal closure instants* (GCI) has been proposed to resolve these issues (Pollard *et al.* 1997). An alternative is to remove the linear phase from the harmonic phase (Stylianou 2001). More recently, phase-aware signal processing solutions have been applied to achieve high quality speech synthesis. For example, to model the amplitude and phase spectra together, Maia proposed using the complex cepstrum (Maia *et al.* 2012). Maia and Stylianou (2014) proposed *complex cepstrum factorization* for acoustic modeling in a statistical parametric synthesizer. Finally, iterative phase estimation in the complex domain was successfully applied for speech synthesis (Maia and Stylianou, 2016). All these findings confirm the importance of phase-aware signal processing to maintain high quality synthetic speech. This requires a proper model for phase information, which cannot be accomplished by spectral magnitude information. Therefore, deeper insights into complex cepstrum signal models could be considered as future directions to study for speech synthesis.

7.3.1.3   Phase-Aware Speech Enhancement for De-Reverberation

While many single-channel speech de-reverberation and enhancement algorithms exist, they mostly rely on estimating the clean signal amplitude and use the noisy observed phase for signal reconstruction. The recent success in phase-aware processing as reviewed in Chapter 4 is expected to be helpful for a speech de-reverberation algorithm in two possible ways: (i) to estimate the clean phase and use it to reconstruct an estimate of the clean signal as input to one of the state-of-the-art de-reverberation algorithms (see Habets 2007, 2010 for an overview), and (ii) a joint ambient noise reduction and reverberation reduction is possible where both spectral phase and amplitude information are enhanced. Another direction to follow is to extend phase processing to multi-channel de-reverberation where each channel is first provided with an enhanced spectral clean phase prior to the de-reverberation algorithm as the post-processor.

7.3.1.4   Iterative Signal Estimation

The iterative approaches applied to recover a signal from its partial STFT information (as some examples, see Griffin and Lim 1984; Hayes *et al.* 1980; Quatieri and Oppenheim 1981) do not take into account any specific model for the underlying signal, for example, speech. In particular, the complex cepstrum domain, mostly unused but of high potential, could be helpful to solve the joint amplitude and phase estimation problem, since for an *autoregressive* (AR) process the two are connected via the Hilbert transform. Recent advances in exploiting the properties of the complex cepstrum (Maia and Stylianou 2016) encourage future effort in combining an iterative signal reconstruction framework and complex domain signal modeling in a unified framework. Combining a signal model for the underlying speech signal together within a multiple input spectrogram inversion (Gunawan and Sen 2010; MISI, discussed in Chapter 5) is expected to outperform the GLA-based approaches reviewed for signal reconstruction. Such a successful combination could be extended to speech enhancement and source separation to synthesize the enhanced and separated outcomes in an iterative framework.

7.3.1.5 More Robust Phase Estimators

In Chapter 3, existing phase estimators for speech signal processing were reviewed and a comparative study was provided. The results of the comparative study show a distinct gap between the performance of phase estimators versus the phase estimation upper bound provided by the clean spectral phase. Also, comparing the outcome of the blind experiments of phase-aware speech enhancement methods in Chapter 4 to the results of the oracle clean spectral phase highlights the relevance of a more accurate estimator for the initial phase with less uncertainty. Given the recent interest shown by the community in phase estimation (Mowlaee and Kulmer 2015a; b; Krawczyk and Gerkmann 2014), the authors believe that significant improved performance in phase estimation of speech in noise is expected in the near future that will eventually allow phase-aware speech enhancement methods to further outperform the conventional amplitude-based speech enhancement approaches.

7.3.1.6 Instrumental Measures in Complex Signal Domain

In Chapter 6, the reliability of some phase-only instrumental metrics was measured and a correlation analysis of the objective measures versus the subjective listening outcome was presented. To take into account distortions or artifacts due to both amplitude and phase spectra, future work should be dedicated to deriving a novel distortion measure working in the complex domain, hence including the distortions due to joint amplitude and phase estimation errors captured by further defining a complex domain distortion measure. Through subjective listening tests and correlation analysis the reliability of such a measure compared to conventional amplitude-only or recent phase-only ones should be assessed. Having access to a reliable instrumental measure for speech quality estimation will serve as an important step in developing new phase-aware speech processing algorithms in different applications.

7.3.1.7 Multi-Channel Speech Processing

This book has addressed phase-aware signal processing with one microphone. The single-channel techniques described for phase estimation (Chapter 3) or phase-aware spectral amplitude estimation for speech enhancement (Chapter 4), source separation (Chapter 5), and speech quality estimation (Chapter 6) could be extended to multiple microphones in a pre-processing or post-processing fashion. As a pre-processing configuration, the clean spectral phase information can be estimated using one phase estimator per microphone channel. Alternatively, a beamformed signal could be post-processed by a single-channel phase-aware speech enhancement method.

In particular, some phase-related cues, for example, *interaural differences in time* (ITD) and *interaural differences in level* (ILD), which are known to be helpful binaural cues, could be improved by applying some of the phase-aware signal processing techniques discussed here. Other relevant applications are target localization and tracking where several acoustic sensors are used.

7.3.2 Other Research Disciplines

Although this book is aimed at speech communication applications, signal processing fields not related to speech communication (e.g, other neighboring domains applied to non-speech signals) could still benefit from the techniques and tools provided here. We briefly point out two groups: non-speech one dimensional signals and processing signals at higher dimensionality than one, explained in the following.

7.3.2.1 Processing Non-Speech Signals

These applications include: music signal processing, audiology in the context of hearing aids, melody extraction, audio source separation, biomedical signal processing including *electromyography* (EMG), *electroencephalography* (EEG), and *electrocardiogram* (ECG) signals. For example, for ECG preprocessing, a linear phase filter is of great importance to remove baseline wander (Miklavcic *et al.*, 2006) and to improve the estimation of the instantaneous phase for different frequency components in EEG (Boyden *et al.* 2008). Other examples are optics (Vyacheslav and Zhu 2003) and radar signal processing (Costantini *et al.* 1999).

7.3.2.2 Processing Signals of Higher Dimensionality Than One

In addition, phase processing for two-dimensional (image: Ying 2006; Ghiglia and Pritt 1998), three-dimensional (video: Wadhwa *et al.*, 2013), or even higher data dimensions (tensors) are active fields of research. *Phase wrapping* is a fundamental problem involved in many research fields such as wireless communication (synthetic aperture radar), interferometry (Goldstein *et al.* 1988), medical imaging (e.g., field mapping in magnetic resonance imaging (MRI); Ying 2006), source localization (Andersen and Jensen 2001), and wavefront distortion measurements in adaptive optics. Phase unwrapping for two-dimensional data has been comprehensively reviewed in Ghiglia and Pritt (1998) and Ying (2006).

7.4 Summary

In this book, we have presented an overview of previous and more recent advances made toward incorporating phase information in processing speech signals. We have described the contributions made by researchers in diverse fields of speech processing applications including speech analysis, speech enhancement, source separation, and speech quality estimation. It is the authors' belief that the current work represents the first step toward a unified study of different perspectives on phase-aware processing, currently scattered in different communities in speech communication. As a result, the current book bridges the gap to share useful knowledge and to improve overall understanding of how phase information could be incorporated in pushing the limits of the conventional state-of-the-art methods. As the recent endeavors are scattered within different subfields of speech processing conducted by many researchers with separate contributions focused on individual research topics, the current book could serve to inform a researcher new to this field with a literature review on the topic of his or her interest, and provide an overview on some adjacent fields, giving some further insights into the positive improvements shown in several speech applications, as provided in the second part of the book.

The book is accompanied by a web page

https://www.spsc.tugraz.at/PhaseLab

that has two purposes. First, it provides the *PhaseLab Toolbox*, which contains the MATLAB® implementations for several of the experiments presented within the book (see the appendix for more details). Second, it provides relevant links to several phase-related events or news which could be useful for active researchers in the field of phase-aware speech signal processing. The main goal of this webpage is to maintain communication with researchers who are interested in the topic of *phase-aware signal processing in speech communication*. The web page also contains several audio demos and listening examples available for perceptual comparison. A mailing list has been collected since the special conference session (Mowlaee *et al.* 2014) to keep the active and interested researchers in this field updated and informed regarding relevant events in the future.

References

Y. Agiomyrgiannakis, *VOCAINE: The Vocoder and Applications in Speech Synthesis*, Proceedings of the IEEE International Conference on Acoustics, Speech and Signal Processing (ICASSP), pp. 4230–4234, 2015.

T. H. Andersen and K. Jensen, *On the Importance of Phase Information in Additive Analysis/Synthesis of Binaural Sounds*, Proceedings of the International Conference on Computer Music, 2001.

E. S. Boyden, Z. M. Anderson, and E. Mishra, *Integrated Transcranial Current Stimulation and Electroencephalography Device*, US Patent App. 12/118,092, 2008.

B. Bozkurt, *Zeros of the z-Transform (ZZT) Representation and Chirp Group Delay Processing for the Analysis of Source and Filter Characteristics of Speech Signals*, PhD Thesis, Faculty Polytechnique De Mons, Belgium, 2005.

M. Costantini, A. Farina, and F. Zirilli, A fast phase unwrapping algorithm for SAR interferometry, *IEEE Transactions on Geoscience and Remote Sensing*, vol. 37, no. 1, pp. 452–460, 1999.

Z. Donglai and K. K. Paliwal, *Product of Power Spectrum and Group Delay Function for Speech Recognition*, Proceedings of IEEE International Conference on Acoustics, Speech, and Signal Processing (ICASSP), vol. 1, pp. 125–128, 2004.

H. Erdogan, J. R. Hershey, S. Watanabe, and J. Le Roux, *Phase-Sensitive and Recognition-Boosted Speech Separation using Deep Recurrent Neural Networks*, Proceedings of the IEEE International Conference on Acoustics, Speech and Signal Processing (ICASSP), 2015.

D. C. Ghiglia and M. D. Pritt, *Two-Dimensional Phase Unwrapping: Theory, Algorithms, and Software*, John Wiley & Sons, 1998.

R. M. Goldstein, H. A. Zebker, and C. L. Werner, Satellite radar interferometry: Two-dimensional phase unwrapping, *Radio Science*, vol. 23, pp. 713–720, 1988.

P. Golik, Z. Tüske, R. Schlüter, and H. Ney, *Convolutional Neural Networks for Acoustic Modeling of Raw Time Signal in LVCSR*, Proceedings of the International Conference on Spoken Language Processing (INTERSPEECH), pp. 26–30, 2015.

D. Griffin and J. Lim, Signal estimation from modified short-time Fourier transform, *IEEE Transactions on Acoustics, Speech, and Signal Processing*, vol. 32, no. 2, pp. 236–243, 1984.

D. Gunawan and D. Sen, Iterative phase estimation for the synthesis of separated sources from single-channel mixtures, *IEEE Signal Processing Letters*, vol. 17, no. 5, pp. 421–424, 2010.

E. A. P. Habets, *Single- and Multi-Microphone Speech Dereverberation using Spectral Enhancement*, PhD Thesis, Technische Universiteit Eindhoven, The Netherlands, 2007.

E. A. P. Habets, Speech dereverberation using statistical reverberation models, in *Speech Dereverberation*, (eds P. A. Naylor and N. D. Gaubitch), Springer, 2010.

M. Hayes, J. Lim, and A. V. Oppenheim, Signal reconstruction from phase or magnitude, *IEEE Transactions on Acoustics, Speech, and Signal Processing*, vol. 28, no. 6, pp. 672–680, 1980.

P. Hedelin, *Phase Compensation in All-Pole Speech Analysis*, Proceedings of the IEEE International Conference on Acoustics, Speech and Signal Processing (ICASSP), pp. 339–342, 1988.

R. M. Hegde, H. A. Murthy, and V. R. R. Gadde, Significance of the modified group delay feature in speech recognition, *IEEE Transactions on Audio, Speech and Language Processing*, vol. 15, no. 1, pp. 190–202, 2007.

H. von Helmholtz, *On the Sensations of Tone as a Physiological Basis for the Theory of Music*, Longmans, Green, 1912.

T. Kleinschmidt, S. Sridharan, and M. Mason, The use of phase in complex spectrum subtraction for robust speech recognition, *Computer Speech & Language*, vol. 25, no. 3, pp. 585–600, 2011.

M. Krawczyk and T. Gerkmann, STFT phase reconstruction in voiced speech for an improved single-channel speech enhancement, *IEEE/ACM Transactions on Audio, Speech, and Language Processing*, vol. 22, no. 12, pp. 1931–1940, 2014.

E. Loweimi, S. M. Ahadi, and T. Drugman, *A New Phase-Based Feature Representation for Robust Speech Recognition*, Proceedings of the IEEE International Conference on Acoustics, Speech and Signal Processing (ICASSP), pp. 7155–7159, 2013.

R. McAulay and T. F. Quatieri, Speech analysis/synthesis based on a sinusoidal representation, *IEEE Transactions on Acoustics, Speech, and Signal Processing*, vol. 34, no. 4, pp. 744–754, 1986.

R. Maia and Y. Stylianou, *Complex Cepstrum Factorization for Statistical Parametric Synthesis*, Proceedings of the IEEE International Conference on Acoustics, Speech and Signal Processing (ICASSP), pp. 3839–3843, 2014.

R. Maia and Y. Stylianou, *Iterative Estimation of Phase Using Complex Cepstrum Representation*, Proceedings of the IEEE International Conference on Acoustics, Speech and Signal Processing (ICASSP), 2016.

R. Maia, M. Akamine, and M. J. F. Gales, *Complex Cepstrum as Phase Information in Statistical Parametric Speech Synthesis*, Proceedings of the IEEE International Conference on Acoustics, Speech and Signal Processing (ICASSP), pp. 4581–4584, 2012.

D. Miklavcic, N. Pavselj, and F. X. Hart, *Wiley Encyclopedia of Biomedical Engineering*, John Wiley & Sons, Inc., 2006.

P. Mowlaee and J. Kulmer, Phase estimation in single-channel speech enhancement: Limits-potential, *IEEE/ACM Transactions on Audio, Speech, and Language Processing*, vol. 23, no. 8, pp. 1283–1294, 2015a.

P. Mowlaee and J. Kulmer, Harmonic phase estimation in single-channel speech enhancement using phase decomposition and SNR information, *IEEE/ACM Transactions on Audio, Speech, and Language Processing*, vol. 23, no. 9, pp. 1521–1532, 2015b.

P. Mowlaee, R. Saeidi, and Y. Stylianou, *Special Session on Phase Importance in Speech Processing*, Proceedings of the International Conference on Spoken Language Processing (INTERSPEECH), pp. 1623–1627, 2014.

P. Mowlaee, R. Saeidi, and Y. Stylianou, *Special Issue on Phase-Aware Signal Processing in Speech Communication*, [Online]. Available: http://www.journals.elsevier.com/speech-communication/call-for-papers/special-issue-on-phase-aware-signal-processing-in-speech-co/, 2015.

K. S. R. Murty and B. Yegnanarayana, Combining evidence from residual phase and MFCC features for speaker recognition, *IEEE Signal Processing Letters*, vol. 13, no. 1, pp. 52–55, 2006.

S. Nakagawa, L. Wang, and S. Ohtsuka, Speaker identification and verification by combining MFCC and phase information, *IEEE Transactions on Acoustics, Speech, and Signal Processing*, vol. 20, no. 4, pp. 1085–1095, 2012.

G. S. Ohm, Über die Definition des Tones, nebst daran geknüfter Theorie der Sirene und ähnlicher tonbildender Vorichtungen, *Journal of Physical Chemistry*, vol. 59, pp. 513–565, 1843.

A. V. Oppenheim, A speech analysis–synthesis system based on homomorphic filtering, *The Journal of the Acoustical Society of America*, vol. 45, no. 2, pp. 458–465, 1969.

H. Pobloth and W. B. Kleijn, *On Phase Perception in Speech*, Proceedings of the IEEE International Conference on Acoustics, Speech and Signal Processing (ICASSP), pp. 29–32, 1999.

H. Pobloth and W. B. Kleijn, Squared error as a measure of perceived phase distortion, *The Journal of the Acoustical Society of America*, vol. 114, no 2, pp. 1081–1094, 2003.

M. P. Pollard, B. M. G. Cheetham, and M. D. Edgington, *Shape Invariant Pitch and Time-Scale Modification of Speech by Variable Order Phase Interpolation*, Proceedings of the IEEE International Conference on Acoustics, Speech and Signal Processing (ICASSP), pp. 919–922, 1997.

T. F. Quatieri, Minimum and mixed-phase speech analysis–synthesis by adaptive homomorphic deconvolution, *IEEE Transactions on Acoustics, Speech, and Signal Processing*, vol. 27, no. 4, pp. 328–335, 1979.

T. F. Quatieri and A. V. Oppenheim, Iterative techniques for minimum phase signal reconstruction from phase or magnitude, *IEEE Transactions on Acoustics, Speech, and Signal Processing*, vol. 29, no. 6, pp. 1187–1193, 1981.

S. O. Sadjadi and J. H. L. Hansen, *Hilbert Envelope Based Features for Robust Speaker Identification under Reverberant Mismatched Conditions*, Proceedings of the IEEE International Conference on Acoustics, Speech and Signal Processing (ICASSP), pp. 5448–5451, 2011.

Y. Stylianou, Applying the harmonic plus noise model in concatenative speech synthesis, *IEEE Transactions on Speech and Audio Processing*, vol. 9, no. 1, pp. 21–29, 2001.

Z. Tüske, P. Golik, R. Schlüter, and H. Ney, *Acoustic Modeling with Deep Neural Networks Using Raw Time Signal for LVCSR*, Proceedings of the International Conference on Spoken Language Processing (INTERSPEECH), pp. 890–894, 2014.

V. Vyacheslav and Y. Zhu, Deterministic phase unwrapping in the presence of noise, *Optics Letters*, vol. 28, no. 22, pp. 2156–2158, 2003.

N. Wadhwa, M. Rubinstein, F. Durand, and W. T. Freeman, Phase-based video motion processing, *ACM Transactions on Graphics* vol. 32, no. 4, 2013.

D. S. Williamson, Y. Wang, and D. Wang, Complex ratio masking for monaural speech separation, *IEEE/ACM Transactions on Audio, Speech, and Language Processing*, vol. 24, no. 3, pp. 483–492, 2016.

L. Ying, Phase unwrapping, in *Wiley Encyclopedia of Biomedical Engineering*, John Wiley & Sons, 2006.

A

MATLAB Toolbox

A.1   Chapter Organization

Due to the strong link between theory and practice in speech communication applications, this book is supplemented with many experiments to demonstrate the usefulness of phase-aware processing in several different applications. In addition, it is the authors' strong belief that having access to the corresponding MATLAB® code is necessary to move the young field of phase-aware speech processing forward. This appendix provides the list of implementations used to produce the results presented in the book. The contents and the corresponding sections where the experiment was used are described. A detailed description of the *PhaseLab Toolbox* is provided.

A.2   PhaseLab Toolbox

We introduce the *PhaseLab Toolbox*, comprising the MATLAB® implementations of selected experiments presented in the course of this book. For each Chapter, the files are organized into two folders:

- *main folder*, containing implementations of the experiments themselves;
- *additional functions*, which are called from within the main files.

A.2.1   MATLAB® Code

One subfolder is dedicated to each chapter; here, the MATLAB® implementations are located together with *readme.txt* and *readme.pdf* files, which provide information on how to use the files available in the *PhaseLab Toolbox*. A list of the files included in the *PhaseLab Toolbox* is shown in Table A.1. For further reference, for each file a description together with the figure or experiment where it is used are also provided in the table. The MATLAB® code files are downloadable as *.rar* archives from https://www.spsc.tugraz .at/PhaseLab.

The web page also provides several audio examples as supplementary material.

Single Channel Phase-Aware Signal Processing in Speech Communication: Theory and Practice, First Edition.
Pejman Mowlaee, Josef Kulmer, Johannes Stahl, and Florian Mayer.
© 2017 John Wiley & Sons, Ltd. Published 2017 by John Wiley & Sons, Ltd.

Table A.1 Filename, description, and experiment number for each MATLAB® implementation used in the book and included in the *PhaseLab Toolbox*.

Filename	Description	Exp./Fig.
Exp1_2.m	Effects of phase modification	Exp. 1.2
Exp1_3.m	Mismatched window experiment	Exp. 1.3
Exp2_1.m	One-dimensional phase unwrapping	Exp. 2.1
Exp2_3.m	Comparative study of group delay spectra	Exp. 2.3
Exp2_5.m	Circular statistics of the spectral phase	Exp. 2.5
Exp2_6.m	Comparative study of phase representations	Exp. 2.6
Exp3_1.m	Monte Carlo simulation: ML versus MAP phase estimator	Exp. 3.1
Exp3_2.m	Monte Carlo simulation: window impact on phase estimation	Exp. 3.2
Exp3_3.m	GLA versus FGLA for phase retrieval	Exp. 3.3
Exp3_4.m	Phase estimation comparative study	Exp. 3.4
Fig4_9.m	Deterministic components and complex coefficients distribution	Fig. 4.9
Exp4_3.m	Sensitivity analysis of phase-aware amplitude estimators	Exp. 4.3
Exp5_1.m	Phase estimation for proof-of-concept signal reconstruction	Exp. 5.1
Exp5_2.m	Comparative study of GLA-based phase reconstruction methods	Exp. 5.2
Exp5_3.m	Phase-aware time frequency masks	Exp. 5.3
Exp5_5.m	Complex matrix factorization (CMF): Figure 5.20	Exp. 5.5
Exp6_2.m	Phase and perceived quality estimation	Exp. 6.2
Exp6_3.m	Phase and speech intelligibility estimation	Exp. 6.3
Exp6_4.m	Evaluating the phase estimation accuracy	Exp. 6.4

A.2.2 Additional Material

Additional material is required when using the code to reproduce the experiments described in the book. For example, speech files selected from the GRID (Cooke *et al.* 2006), SiSEC (Araki *et al.* 2012), or TIMIT (Garofolo *et al.* 1993) databases need to be acquired separately. Also, some experiments require access to other speech processing toolboxes, including COVAREP (Degottex *et al.* 2014), VOICEBOX (Brookes *et al.* 2005), CircStat (Barens 2009), and CMF Toolbox (King and Atlas 2012). For performance evaluation in speech enhancement, perceptual evaluation of speech quality (PESQ; Rix *et al.* 2001) and short-time objective intelligibility measure (STOI; Taal *et al.* 2011) software is required. To quantify the source separation performance, the blind source separation evaluation (BSS EVAL; Vincent *et al.* 2006) is required.

References

S. Araki, F. Nesta, E. Vincent, Z. Koldovský, G. Nolte, A. Ziehe, and A. Benichoux, *The 2011 Signal Separation Evaluation Campaign (SiSEC2011): Audio Source Separation*, Proceedings of the International Conference on Latent Variable Analysis and Signal Separation (LVA/ICA), pp. 414–422, 2012.

P. Barens, CircStat: A MATLAB toolbox for circular statistics, *Journal of Statistical Software*, vol 31, no. 10, pp. 1–21, 2009.

M. Brookes *et al.*, VOICEBOX: Speech Processing Toolbox for MATLAB, [Online], http://www.ee.ic.ac.uk/hp/staff/dmb/voicebox/voicebox.html, 2005.

M. Cooke, J. Barker, S. Cunningham, and X. Shao, An audio-visual corpus for speech perception and automatic speech recognition, *The Journal of the Acoustical Society of America*, vol. 120, pp. 2421–2424, 2006.

G. Degottex, J. Kane, T. Drugman, T. Raitio, and S. Scherer, *COVAREP: A Collaborative Voice Analysis Repository for Speech Technologies*, Proceedings of the IEEE International Conference on Acoustics, Speech and Signal Processing (ICASSP), pp. 960–964, 2014.

J. S. Garofolo, L. F. Lamel, W. M. Fisher, J. G. Fiscus, D. S. Pallett, and N. L. Dahlgren, *DARPA TIMIT Acoustic Phonetic Continuous Speech Corpus CDROM*, National Institute of Standards and Technology (NIST), 1993.

B. King and L. Atlas, *Complex Matrix Factorization Toolbox Version 1.0 for MATLAB*, [Online], https://sites.google.com/a/uw.edu/isdl/projects/cmf-toolbox, University of Washington, 2012.

A. W. Rix, J. G. Beerends, M. P. Hollier, and A. P. Hekstra, *Perceptual Evaluation of Speech Quality (PESQ): A New Method for Speech Quality Assessment of Telephone Networks and Codecs*, Proceedings of the IEEE International Conference on Acoustics, Speech and Signal Processing (ICASSP), vol. 2, pp. 749–752, 2001.

C. H. Taal, R. C. Hendriks, R. Heusdens, and J. Jensen, An algorithm for intelligibility prediction of time–frequency weighted noisy speech, *IEEE Transactions on Audio, Speech, and Language Processing*, vol. 19, no. 7, pp. 2125–2136, 2011.

E. Vincent, R. Gribonval, and C. Févotte, Performance measurement in blind audio source separation, *IEEE Transactions on Audio, Speech, and Language Processing*, vol. 14, no. 4, pp. 1462–1469, 2006.

Index

Single Channel Phase-Aware Signal Processing in Speech Communication: Theory and Practice, First Edition.
Pejman Mowlaee, Josef Kulmer, Johannes Stahl, and Florian Mayer.
© 2017 John Wiley & Sons, Ltd. Published 2017 by John Wiley & Sons, Ltd.